How to talk about Mathematics. Wie man über Mathematik spricht

Sixtus Kage

How to talk about Mathematics. Wie man über Mathematik spricht.

A Handbook for People who Learn, Teach or Apply Mathematics.
Ein Handbuch für Leute, die Mathematik lernen, lehren oder anwenden.

First Edition. Erste Auflage.

2025

All words and phrases and their translations are presented to the best of my knowledge. However, the accuracy of the material presented here and its suitability for your specific communicative purposes is not guaranteed. This work constitutes a research effort to find out and document the current, accepted usage. The latter might also be subject to change over time or to regional variations. In some cases, there may also be differences between scientific discliplines. You are strongly advised to conduct your own research based on the material presented.
Alle Wörter und Redewendungen und ihre Übersetzungen werden nach meinem besten Wissen dargestellt. Jedoch wird die Genauigkeit des dargestellten Materials und seine Eignung für Ihre speziellen kommunikativen Zwecke nicht garantiert. Diese Arbeit stellt eine Bemühung dar, den aktuellen Stand des Sprachgebrauchs zu erforschen und zu dokumentieren. Letzterer könnte auch zeitlichen und regionalen Veränderungen unterliegen. In einigen Fällen könnte es auch Unterschiede zwischen verschiedenen Disziplinen der Wissenschaft geben. Es wird Ihnen dringend empfohlen, eigene Untersuchungen auf der Grundlage des hier vorgelegten Materials anzustellen.

All figures were drawn by Sixtus Kage.
Alle Abbildungen wurden von Sixtus Kage gezeichnet.

Pronunciation: I have read this book out, page by page, in English and in German. This service is free of charge, not included in the price of the book and can be modified or terminated any time.
Ausspräche: Ich habe dieses Buch seitenweise vorgelesen, auf Deutsch und auf Englisch. Dieser Service ist kostenlos, ist nicht im Buchpreis enthalten und kann jederzeit modifiziert oder beendet werden.
See: / Siehe: https://www.youtube.com/@SixtusKage

Corrections and suggestions are always welcome.
Korrekturen und Vorschläge sind immer willkommen.

Sixtus Kage
Postfach 801066
81610 München
Deutschland

Sixtus Kage
PO Box 801066
81610 Munich
Germany

e-mail: sixtus.kage@web.de

Bibliografische Information der Deutschen Nationalbibliothek: Die Deutsche Nationalbibliothek verzeichnet diese Publikation in der Deutschen Nationalbibliografie; detaillierte bibliografische Daten sind im Internet über dnb.dnb.de abrufbar.

Verlag: BoD · Books on Demand GmbH, Überseering 33, 22297 Hamburg, bod@bod.de
Druck: Libri Plureos GmbH, Friedensallee 273, 22763 Hamburg

ISBN: 978-3-8192-7873-0

B Der Zweck dieses Buches. / The Purpose of this Book

B1 Zielgruppe / Target Group

This book is intended to assist people who have to learn, to teach or to apply mathematics while using a foreign language, in this case English or German. These people need the language of describing and explaining mathematical concepts, symbols, and procedures.

Dieses Buch soll Leute unterstützen, die Mathematik lernen, anwenden oder lehren müssen, während sie sich einer fremden Sprache bedienen, in diesem Falle Englisch oder Deutsch. Diese Leute benötigen die Sprache der Beschreibung und Erklärung mathematischer Begriffe, Symbole und Prozeduren.

B2 Linguistic Background / Linguistischer Hintergrund

My concern in this book is not mainly the technical terms, but also the words in between. That these pose the real problem was recognized long ago by John Maddox, the Editor-in-Chief of the prestigious journal *Nature*. "It is not in the vocabulary of science that the difficulty lies, but in the general vocabulary that holds it together" (Maddox 1992: 475). In this respect, dictionaries of mathematics are of limited use. The same applies to the pronunciation of symbols.
However, professor Arens and his colleagues have authored two substantial German books on university mathematics which contain everything you should know in the first two years. One of the books is written for students of natural science or engineering (Arens, Hettlich e.a. 2022), the other one for students of pure mathematics (Arens, Busam e.a. 2022). To the best of my knowledge, these two books are the first of their kind to contain comprehensive lists of mathematical symbols and their pronunciation (Arens, Hettlich e.a. 2022; 1645-1653; Arens, Busam e.a. 2022: 1153-1170). In one of the books, the pronunciation is not only given in German but also in English (Arens, Busam e.a. 2022: 1153-1170). I have found two books which provide a similar service on the level of school mathematics (Städtler & Städtler 1966; Lewisch & Posamentier 2000).
The booklet presented here contains sentences which describe a mathematical context. We speak a foreign languge, but also our first language, by drawing upon a large number of sentences and sentence fragments in our memory. Therefore, any new expressions we need to learn should be introduced in sentences. I am going to describe the details in a psycholinguistic publication.

Meine Sorge in diesem Buch gilt nicht nur den Fachausdrücken, sondern auch den Wörtern dazwischen. Diese stellen das eigentliche Problem dar, wie vor langer Zeit John Maddox erkannte, der Chefredakteur der angesehenen Fachzeitschrift Nature: „Es ist nicht der Wortschatz der Wissenschaft, in welchem die Schwierigkeit liegt, sondern in dem allgemeinen Vokabular, das diesen zusammenhält" (Maddox 1992: 475, überstzt durch SK). Wörterbücher der Mathematik sind in dieser Hinsicht von begrenztem Nutzen. Dasselbe gilt im Hinblick auf die Aussprache von Symbolen.
Jedoch haben Professor Arens und seine Kollegen zwei umfangreiche deutschsprachige Bücher der Universitätsmathematik verfasst, die alles enthalten, was man in den ersten zwei Studienjahren wissen sollte. Eines der Bücher wurde für Studierende der Natur-

und Ingenieurwissenschaften geschrieben (Arens, Hettlich e.a. 2022), das andere für Studierende der reinen Mathematik (Arens, Busam e.a. 2022). Nach meinem besten Wissen handelt es sich um die ersten Bücher dieser Art, die umfassende Listen mathematischer Symbole und ihrer Aussprache enthalten (Arens, Hettlich e.a. 2022: 1645-1653; Arens, Busam e.a. 2022: 1153-1170). In einem der Bücher wird die Aussprache nicht nur auf Deutsch angegeben, sondern auch auf Englisch (Arens, Busam e.a. 2022: 1153-1170). Ich habe zwei Bücher gefunden, die ähnliche Dienste auf der Ebene der Schulmathematik leisten (Städtler & Städtler 1966; Lewisch & Posamentier 2000).
Das vorliegende Büchlein enthält Sätze, die einen mathematischen Zusammenhang beschreiben. Denn wir sprechen eine Fremdsprache, aber auch unsere Erstsprache, indem wir auf eine große Zahl von Sätzen und Satzbruchstücken in unserem Gedächtnis zurückgreifen. Näheres dazu werde ich in einer psycholinguistischen Veröffentlichung schreiben.

B3 Vorgehensweise / Method

For every topic, several roughly equivalent sentences are introduced. Thus, the readers can see the wide variety of expressions suitable for the same purpose. The readers are invited to look for other ways of expressing the same things while stuydying books on mathematics and its applications. Please let me know good sources you might find by e-mail. I am also going to continue my search to be able to expand my lists in the second edition. Furthermore, the google book search (books.google.com) may be consulted. The phrases you and I have found should be used as a search string in quotation marks. If only older books are reported by the book search, then we can conclude that the phrases tested are a bit outdated. The number of books listed by the book search for each phrase contains some information about its popularity. You may also decide, as I have done, to assemble a long list of equivalent phrases to decide which one best conforms to your personal style. You might also like to consult the books I quote from to learn a bit of mathematics. Sources are listed whenever possible so that you do not have to accept the phrases on my authority alone. Furthermore, the book titles allow you to see in which branch of science the phrases are used.
Thus, long quotes are also to be understood as examples of good practice and, consequently, as recommendations of the books cited (e.g.: Stewart, Redlin & Watson 2002; Mortimer & Müller 2015). In the case of older books I recommend this way, the publishers might be motivated to offer a reprint as an e-book.
Lastly, I am going to provide the pronunciation of all sentences by reading this booklet out to you, page by page, in German and in English. Step-by-step, every page is going to be posted to my YouTube channel (https://www.youtube.com/@SixtusKage) as a short video clip twice, one clip in German, the other one in English.
Admittedly, calling this booklet a “handbook” is somewhat of an exaggeration, much like calling a seedling a “tree”. But without the seedling, there will be no tree. This book is a work in progress. It is going to be expanded regularly. The next subjects to be included are, among others, contingency tables, linear regression and correlation, logistic regression, geometry, and topographical surveying. Readers are invited to submit additional proposals.

Zu jedem Thema werde einige, grob gleichwertige Sätze eingeführt. Damit können die Leserinnen und Leser die große Vielfalt der Ausdrücke sehen, die für denselben Zweck geeignet sind. Leserinnen und Leser sind eingeladen, nach anderen Möglichkeiten zu suchen, dasselbe auszudrücken. Dies könnte sich ergeben, wenn sie Bücher zur Mathematik und ihren Anwendungen studieren. Teilen Sie mir bitte gute Quellen per e-mail mit. Ich werde auch meine Suche fortsetzen, um meine Listen in der zweiten Auflage erweitern zu können. Darüberhinaus könnte die Google-Buchsuche (books.google.com) zurate gezogen werden. Redewendungen, die Sie und ich gefunden haben, sollten als Such-Zeichenkette in Anführungszeichen eingegeben werden. Falls nur ältere Bücher durch die Buchsuche gemeldet werden, können wir schließen, dass die gesteten Redewendungen etwas veraltet sind. Die Zahl der zu jeder Redewendung aufgeführten Bücher enthält Informationen über die Popularität der Redewendungen. Sie können sich, wie ich, auch dazu entschließen, eine lange Liste gleichwertiger Redewendungen zu erstellen, um zu entscheiden, welche am besten Ihrem Stil entspricht. Sie könnten sich auch entschließen, die von mir zitierten Bücher zu konsultieren, um etwas Mathematik zu lernen. Quelle werden aufgeführt, wo immer möglich, damit Sie die Redwendungen nicht alleine aufgrund meiner Autorität annehmen müssen. Darüberhinaus erlauben es Ihnen die Buchtitel zu sehen, in welchem Zweig der Wissenschaft die zitierten Redewendungen benutzt werden.

Somit sind lange Zitate als Empfehlungen guter Praxis und, folglich als Empfehlungen der zitierten Bücher zu verstehen (e.g.: Stewart, Redlin & Watson 2002; Mortimer & Müller 2015). Im Falle älterer Bücher, die ich auf solche Art empfehle, könnten die Verlage motiviert sein, einen Nachdruck als e-Book anzubieten.

Schließlich werde ich die Aussprache aller Sätze zur Verfügung stellen, indem ich Ihnen dieses Büchlein vorlese, Seite für Seite, auf Deutsch und auf Englisch. Schrittweise wird jede Seite doppelt auf https://www.youtube.com/@SixtusKage, meinem Kanal auf YouTube, als kurzer Videofilm veröffentlicht werden, ein Film auf Deutsch, der andere auf Englisch.

Zugegeben, dieses Büchlein ein “Handbuch“ zu nennen, ist etwas übertrieben, so wie etwa einen Setzling „Baum“ zu nennen. Aber ohne Setzling wird es keinen Baum geben. Das Buch ist ein Projekt im Werden. Es wird regelmäßig erweitert werden. Als nächste Themen werden unter anderem Vierfeldertafeln, lineare Regression und Korrelation, logistische Regression, Geometrie und Vermessungswesen berücksichtigt. Leserinnen und Leser sind eingeladen, weitere Vorschläge zu unterbreiten.

B4 Hints for Students / Hinweise für Studierende

In Germany, university students have always tended to overestimate their level of preparation when it comes to the required mathematics: "They think they are mediocre, but they are much worse" (Kern 1993: 9, translated by SK). Nothing seems to have changed in the last three decades: "The accumulated experience of a six-person-team of authors might possibly suffice to support an almost incredible assertion: For many freshmen, the most important sources of mistakes in their first year of studying mathematics are a lack of security in dealing with fractions, with power notation and with solving equations!"(Arens, Hettlich e.a. 2022: 52; translated by SK).
The topics I have selected for this booklet so far, e.g., dealing with fractions, power laws and the odds ratio, might seem odd. But this is what I had to go through with, e.g., the medical students I tutored. I helped them both to survive their physics laboratory courses and to understand the contingency tables and the logistic regression they needed to evaluate the data for their doctoral theses.
If you are in the ninth, tenth or eleventh grade and a bit concerned about your proficiency in mathematics (or a bit bored at school), three books I have used could be most helpful to learn some important methods (Kreul & Ziebarth 2016; Stewart, Redlin & Watson 2002; Leupold e.a. 1976). Two books by Warren Esty (Esty 1997; Esty 2012) will enable you to understand the language of mathematics more thoroughly. If you (or your children) need to learn some basic mathematics in English, you might like the series of books written by McSeveny and colleagues (Volume 1: McSeveny, Conway, Wilkes & Smith 2014).

In Deutschland haben Universitätsstudenten immer schon das Niveau ihrer Vorkenntnisse im Hinblick auf die benötigte Mathematik überschätzt: „Sie halten sich für mittelmäßig, sind jedoch viel schlechter" (Kern 1993: 9). In den letzen drei Jahrzehnten scheint sich nichts geändert zu haben: „Die gesammelte Erfahrung eines sechsköpfigen Autorenteams reicht aber vielleicht aus, um eine nahezu unglaublich klingende Behauptung zu untermauern: Mangelnde Sicherheit im Bruchrechnen, beim Umgang mit Potenzen und beim Lösen von Gleichungen ist für viele Studierende die größte Fehlerquelle in den ersten beiden Semestern ihrer Mathematikausbildung!"(Arens, Hettlich e.a. 2022: 52).
Die bisherige Auswahl an Themen, z.B. Bruchrechnen, Potenzrechenregeln und das Wettverhälnis, könnte seltsam erscheinen. Aber genau diese Themen hatte ich z.B. den Medizinstudentinnen zu erklären, deren Nachhilfelehrer ich war. Ich half ihnen sowohl dabei, das Physikpraktikum zu überstehen, als auch dabei, die Vierfeldertafel und die Logistische Regression zu verstehen, mit welchen die Daten ihrer Doktorarbeiten auszuwerten waren.
Wenn Sie in der Klasse 9, 10 oder 11 sind und etwas besorgt über Ihren Leistungsstand in Mathematik sind (oder in der Schule etwas gelangweilt sind), könnten drei Bücher, die ich benutzt habe, sehr hilfreich sein, um einige wichtige Methoden zu lernen (Kreul & Ziebarth 2016; Stewart, Redlin & Watson 2002; Leupold e.a. 1976). Zwei Bücher von Warren Esty (Esty 1997; Esty 2012) werden Ihnen ermöglichen, die Sprache der Mathematik gründlicher zu verstehen. Falls Sie (oder Ihre Kinder) grundlegende Mathematik auf Englisch lernen müssen, könnte Ihnen die von McSeveny und Kollegen geschriebene Buchreihe (Band 1: McSeveny, Conway, Wilkes & Smith 2014) gefallen.

C Themen / Topics

AD1 Addition 1 / Addition 1

AD1_1_e: “Ten plus five equals fifteen.”

AD1_2_e: “The sum of ten and five equals fifteen.”

AD1_3_e: “If you add five to ten, you get fifteen.”

AD1_4_e: “Adding five to ten yields fifteen.”

AD1_5_e: “The sum of the numbers ten and five is fifteen.”

AD1_6_e: “The number ten is the first addend, five is the second, 15 is the sum.”

AD1_7_e: “x plus y equals z.”

AD1_8_e: “The sum of x and y is z.”

AD1_9_e: “If you add x and y, you get z.”

AD1_10_e: “Adding x and y yields z.”

AD1_11_e: “The sum of the numbers x and y is the number z.”

AD1_12_e: “x is the first addend, y is the second addend, z is the sum.”

$10 + 5 = 15$ $x + y = z$

AD1_1_d: „Zehn plus fünf gleich fünfzehn.”

AD1_2_d: „Die Summe aus zehn und fünf ist fünfzehn.”

AD1_3_d: „Wenn man fünf zu zehn addiert, erhält man fünfzehn.”

AD1_4_d: „Die Addition von zehn und fünf ergibt fünfzehn.”

AD1_5_d: „Die Summe der Zahlen zehn und fünf ist fünfzehn.“

AD1_6_d: „Die Zahl zehn ist der erste Summand, die Zahl fünf ist der zweite Summand, die Zahl fünfzehn ist die Summe.

AD1_7_d: „x plus y gleich z.“

AD1_8_d: „Die Summe von x und y ist z.“

AD1_9_d: „Wenn man x und y addiert, erhält man z.“

AD1_10_d: „Die Addition von x und y liefert z.“

AD1_11_d: „Die Summe der Zahlen x und y ist die Zahl z.“

AD1_12_d: „x ist der erste Summand, y ist der zweite Summand, z ist die Summe.“

ST1 Subtraction 1 / Subtraktion 2

ST1_1_e: "Fifteen minus five equals ten."

ST1_2_e: "The difference of fifteen and five is ten."

ST1_3_e: "If you subtract five from fifteen, you get ten."

ST1_4_e: "Subtracting five from fifteen yields ten."

ST1_5_e: "The difference of the numbers fifteen and five is ten."

ST1_6_e: "The number fifteen is the minuend, five is the subtrahend, 15 is the difference."

ST1_7_e: "x minus y equals z."

ST1_8_e: "The difference of x and y is z."

ST1_9_e: "If you subtract y from x, you get z."

ST1_10_e: "Subtracting y from x yields z."

ST1_11_e: "The difference of the numbers x and y is the number z.."

ST1_12_e: "x is the minuend, y is the subtrahend, z is the difference."

$15 - 5 = 10$ $x - y = z$

ST1_1_d: „Fünfzehn minus fünf gleich zehn."

ST1_2_d: „Die Differenz von fünfzehn und fünf ist zehn."

ST1_3_d: „Wenn man fünf von fünfzehn subtrahiert, erhält man zehn."

ST1_4_d: „Die Subtraktion der Zahl fünf von fünfzehn ergibt zehn."

ST1_5_d: „Die Differenz der Zahlen fünfzehn und fünf ist zehn."

ST1_6_d: „Die Zahl fünfzehn ist der Minuend, fünf ist der Subtrahend und zehn ist die Differenz."

ST1_7_d: „x minus y gleich z."

ST1_8_d: „Die Differenz von x und y ist z."

ST1_9_d: „Wenn man y von x subtrahiert, erhält man z."

ST1_10_d: „Die Subtraktion der Zahl y von der Zahl x liefert z."

ST1_11_d: „Die Differenz Zahlen x und y ist die Zahl z."

ST1_12_d: „x ist der Minuend, y ist der Subtrahend, z ist die Differenz."

MU1 Multiplication 1 / Multiplikation 1

MU1_1_e: “Three times four equals twelve.”

MU1_2_e: “The product of three and four is twelve.”

MU1_3_e: “If you multipy three and four, you get twelve.”

MU1_4_e: “Multiplying three by four yields twelve.”

MU1_5_e: “Multiplying three and four yields twelve.”

MU1_6_e: “The product of the numbers three and four ist twelve.”

MU1_7_e: “The number three is the first factor, four is the second factor, 12 is the product.”

MU1_8_e: “x times y equals z.”

MU1_9_e: “The product of x and y is z.”

MU1_10_e: “If you multiply x and y, you get z.”

MU1_11_e: “If you multiply x by y, you get z.”

MU1_12_e: “The multiplication of x and y yields z.”

MU1_13_e: “The product of the numbers x and y is the number z..”

MU1_14_e: “x is the first factor, y is the second factor, z is the difference.”

$3 \times 4 = 12$ $x \times y = z$ $3 \cdot\cdot 4 = 12$ $x \cdot y = z$

MU1_1_d: „Drei mal vier gleich zwölf.”

MU1_2_d: „Das Produkt von drei und vier ist zwölf.”

MU1_3_d: „Wenn man drei und vier multipliziert, erhält man zwölf.”

MU1_4_d: „Wenn man drei mit vier multipliziert, erhält man zwölf.”

MU1_5_d: „Die Multiplikation der Zahlen drei und vier ergibt zwölf.”

MU1_6_d: „Das Produkt der Zahlen drei und vier ist zwölf.“

MU1_7_d: „Die Zahl drei ist der erste Faktor, vier ist der zweite Faktor, zwölf ist das Produkt.“

MU1_8_d: „x mal y gleich z.“

MU1_9_d: „Die Produkt von x und y ist z.“

MU1_10_d: „Wenn man x und y multipliziert, erhält man z.“

MU1_11_d: „Wenn man x mit y multipliziert, erhält man z.“

MU1_12_d: „Das Produkt der Zahlen x und y ist z.“

MU1_13_d: „x ist der erste Faktor, y ist der zweite Faktor, z ist das Produkt.“

DV1 Division 1 / Division 1

DV1_1_e: “Fifteen divided by three equals five.”

DV1_2_e: “The quotient of fifteen and three is five.”

DV1_3_e: “If you divide fifteen by three, you get five.”

DV1_4_e: “Dividing fifteen by three yields five.”

DV1_5_e: “If you divide fifteen by three, you get five.”

DV1_6_e: “The number fifteen is the dividend, three ist the divisor, 15 is the quotient.”

DV1_7_e: “Fifteen over three equals five.” (a division seen as a fraction)

DV1_8_e: “x divided by y equals z.”

DV1_9_e: “The quotient of x and y is z.”

DV1_10_e: “If you divide x by y, you get z.”

DV1_11_e: “Dividing x by y yields z.”

DV1_12_e: “The quotient of the numbers x and y is the number z.”

DV1_13_e: “x is the dividend, y is the divisor, z is the quotient

DV1_14_e: “x over y equals z.”

$15 \div 3 = 5$ $\quad$ $x \div y = z$ $\quad$ $15 : 3 = 5$ $\quad$ $x : y = z$ $\quad$ $\frac{15}{3} = 5$ $\quad$ $\frac{x}{y} = z$

DV1_1_d: „Fünfzehn dividiert durch drei gleich fünf.”

DV1_2_d: „Der Quotient von fünfzehn und drei ist fünf.”

DV1_3_d: „Wenn man fünfzehn durch drei dividiert, erhält man fünf.”

DV1_4_d: „Das Teilen von 15 durch drei liefert fünf.”

DV1_5_d: „Wenn man fünfzehn durch drei teilt, erhält man fünf.“

DV1_6_d: „Die Zahl fünfzehn ist der Dividend, drei ist der Divisor und fünf ist der Quotient.“

DV1_7_d: „Fünfzehn durch drei gleich fünf.“ (eine Division als Bruch angesehen)

DV1_8_d: „x dividiert durch y gleich z.“

DV1_9_d: „Die Quotient von x und y ist z.“

DV1_10_d: „Wenn man x durch y dividiert, erhält man z.“

DV1_11_d: „Die Division der Zahl x durch y liefert z.“

DV1_12_d: „Der Quotient der Zahlen x und y ist die Zahl z.“

DV1_13_d: „x ist der Dividend, y is der Divisor, z ist der Quotient.“

DV1_13_e: „x durch y gleich z.”

AB1: Absolute Value and Number Line / Betrag und Zahlenstrahl

AB1_1_e: If x is a real number, $|x|$, the "absolute value of *x*"(Beck & Geoghegan 2010: 176), has thefollowing definition: $|x|$= x für x>0, $|x|$=0 für x=0, $|x|$=-x für x<0. Read: "The absolute value of x equal x, if x is positive. The absolut evalue of x equals zero, if x is zero. The absolute value of x equals negative x if x is negative."

AB1_2_e: "The absolute vaule of a number *a*, denoted by $|a|$, is the distance from *a* to 0 on the real number line" (Stewart, Redlin & Watson 2002: 10, boldface removed).

AB1_3_e: "Graphically, the absolute value $|a|$ describes the *distance* of the point belonging to the number *a* from zero on the number line " (Kreul & Ziebarth 2016: 145, italivs in the source, translated by SK).

AB1_4_e: The absolute value of x minus y, $|x-y|$, is the "distance from *x* to *y*"(Beck & Geoghegan 2010: 176) on the number line. In the figure, the distance from -6 to 2 equals $|(-6)-2|=|-8|=8$ or $|2-(-6)|=|2+6|=|8|=8$.

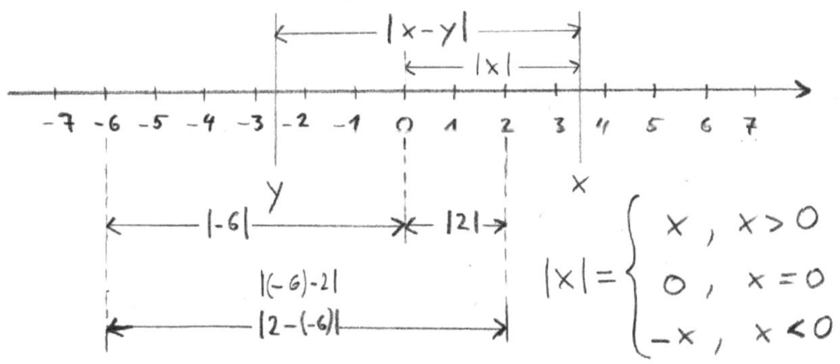

Figure: Absolute value and number line

Abbildung: Betrag und Zahlenstrahl

AB1_1_d: $|x|$, der Betrag einer reellen Zahl x, ist folgendermaßen definiert: $|x|$= x für x>0, $|x|$=0 für x=0, $|x|$=-x für x<0. Lies: „Betrag von x gleich x für x größer null, Betrag von x gleich null für x gleich null, Betrag von x gleich minus x für x kleiner null"

AB1_2_d: „Der Betrag einer Zahl *a*, bezeichnet durch $|a|$, ist der Abstand von *a* zur 0 auf der reellen Zahlengraden " (Stewart, Redlin & Watson 2002: 10, Fettdruck entfernt, kursiv im Original, überstzt von SK).

AB1_3_d: „Anschaulich bescheibt der Betrag $|a|$ den *Abstand* des zu einer Zahl *a* gehörenden Punktes vom Nullpunkt der Zahlengeraden" (Kreul & Ziebarth 2016: 145, kursiv im Original).

AB1_4_d: Der Betrag von x minus y, $|x-y|$, is the "Abstand von *x* und *y*"(Beck & Geoghegan 2010: 176) auf der Zahlengeraden. In der Abbildung ist der Abstand von -6 und 2 gleich $|(-6)-2|=|-8|=8$ oder $|2-(-6)|=|2+6|=|8|=8$.

FR1: Bruchrechnung 1 / Fractions 1

FR_1_e: A fraction (or fractional expression) consists of a numerator, a fraction bar, and a denominator. The denominator must not be 0. "a over b" or "a divided by b"

FR1_2_e: Fractions which have the same denominator are added by adding the numerators and keeping the denominator.

FR1_3_e: „The *golden rule for fractions* says you can multiply or divide the numerator and denominator by the same number (other than 0) without changing its value" (Herman 2020: 22, italics in the source).

FR1_4_e: "A fraction is in its lowest terms if the numerator and denominator have no common factor" (Herman 2020: 22). Here, 4/13 is the fraction in its lowest terms.

FR1_5_e: If you multiply a fraction by a number, you multiply only the numerator by this number and leave the denominator as it is.

FR1_6_e: You reduce a fraction to higher (lower) terms by multiplying (dividing) the numerator and the denominator by the same number. The value of the fraction does not change.

$$\frac{a}{b}=\frac{NUMERATOR}{DENOMINATOR}=\frac{ZÄHLER}{NENNER} \qquad \frac{a}{c}+\frac{b}{c}=\frac{a+b}{c},\frac{4}{11}+\frac{5}{11}=\frac{9}{11}$$

$$\frac{8}{26}=\frac{8\div 2}{26\div 2}=\frac{4}{13}=\frac{4\cdot 3}{13\cdot 3}=\frac{12}{39},\frac{x}{y}=\frac{x\cdot z}{y\cdot z} \qquad 3\cdot\frac{4}{13}=\frac{3\cdot 4}{13}=\frac{12}{13},z\cdot\frac{x}{y}=\frac{z\cdot x}{y}$$

FR1_1_d: Ein Bruch (oder Bruchterm) besteht aus einem Zähler, einem Bruchstrich und einem Nenner. Der Nenner darf nicht 0 sein. „a durch b" oder „a geteilt durch b"

FR1_2_d: Brüche, die denselben Nenner haben, werden addiert, indem man die Zähler addiert und den Nenner beibehält (siehe auch Kreul & Ziebarth 2016: 52).

FR1_3_d: „Die *Goldene Regel für Brüche* besagt, dass man den Zähler und den Nenner mit derselben Zahl (außer 0) multiplizieren oder durch dieselbe Zahl dividieren kann, ohne den Wert des Bruches zu ändern "(Herman 2020: 22, kursiv im Original, übersetzt von SK).

FR1_4_d: „Ein Bruch ist vollständig gekürzt, falls der Zähler und der Nenner keinen gemeinsamen Faktor haben" (Herman 2020: 22, übersetzt von SK). Hier ist 4/13 der vollständig gekürzte Bruch.

FR1_5_d: Ein Bruch wird mit einer Zahl multipliziert, indem man nur den Zähler mit dieser Zahl multipliziert und den Nenner unverändert lässt.

FR1_6_d: Ein Bruch wird mit einer Zahl erweitert (gekürzt), indem man Zähler und Nenner mit dieser Zahl multipliziert (durch diese Zahl dividiert). Der Wert des Bruches ändert sich nicht.

[Die auf Deutsch geschriebenen Sätze folgen teilweise einem Arbeitsblatt, das mein Mathematiklehrer H.-J. K. 1973 ausgeteilt hat.]

FR2: Bruchrechnung 2 / Fractions 2

FR2_1_e: Fractions which have different denominators are added or subtracted by finding a common denominator first. This can be done by reducing the fractions to higher or lower terms.

FR2_2_e: "To write a sum or difference of fractional expressions as a single fraction, we bring them to a common denominator" (Stewart, Redlin & Watson 2002: 704).

FR2_3_e: "[T]o multiply two fractions, we multiply their numerators and multiply their denominators" (Stewart, Redlin & Watson 2002: 36).

FR2_4_e: To invert a fraction, exchange the numerator and the denominator.

FR2_5_e: "[T]o divide a fraction by another fraction we invert the divisor and multiply" (Stewart, Redlin & Watson 2002: 37).

$$\frac{1}{x}+\frac{1}{y}=\frac{1\cdot y}{x\cdot y}+\frac{1\cdot x}{y\cdot x}=\frac{y+x}{x\cdot y} \qquad \frac{1}{3}+\frac{1}{5}=\frac{1\cdot 5}{3\cdot 5}+\frac{1\cdot 3}{5\cdot 3}=\frac{5+3}{3\cdot 5}=\frac{8}{15} \qquad \frac{3}{7}\cdot\frac{2}{5}=\frac{3\cdot 2}{7\cdot 5}=\frac{6}{35}$$

$$\frac{1}{\left(\frac{3}{4}\right)}=\frac{4}{3} \qquad \frac{1}{\left(\frac{x}{y}\right)}=\frac{y}{x} \qquad \frac{3}{4}\div\frac{7}{5}=\frac{3}{4}\cdot\frac{5}{7}=\frac{3\cdot 5}{4\cdot 7}=\frac{15}{28} \qquad \frac{x}{y}\div\frac{a}{b}=\frac{x}{y}\cdot\frac{b}{a}=\frac{x\cdot b}{y\cdot a}$$

FR2_1_d: Brüche mit unterschiedlichen Nennern werden addiert oder subtrahiert, indem man zunächst einen gemeinsamen Nenner findet. Dies geschieht durch Erweitern oder Kürzen.

FR2_2_d: "Um eine Summe oder Differenz von Bruchtermen als einen einzigen Bruch zu scheiben, bringen wir diese auf einen gemeinsamen Nenner" (Stewart, Redlin & Watson 2002: 704, übersetzt von SK).

FR2_3_d: „Brüche werden multipliziert, indem man die Zähler und die Nenner jeweils miteinander multipliziert" (Kreul & Ziebarth 2016: 61).

FR2_4_d: Man bildet den Kehrwert eines Bruches, indem man Zähler und Nenner vertauscht.

FR2_5_d: Um einen Bruch durch eine zweiten Bruch zu dividieren, multipliziert man den ersten Bruchs (den Dividenden) mit dem Kehrwert des zweiten (des Divisors).

DT1 Distributive Property 1 / Distributivgesetz 1

DT1_1_e: If you multiply a sum by a number, you can multiply each term in the sum by this number and add the results: a×(b+c)=a×b+a×c.
This is also called "expanding the bracket".
Read: „a times left bracket b plus c right bracket equals a times b plus a times c"

DT1_2_e: The reverse process is called "factoring out a": a×b+a×c= a×(b+c)

DT1_3_e: If you divide a sum by a number, you can divide each term in the sum by the number and add the results: $\frac{a+b}{c}=\frac{a}{c}+\frac{b}{c}$. E.g., $\frac{104}{4}=\frac{100+4}{4}=\frac{100}{4}+\frac{4}{4}=25+1=26$
Read: "a plus b [pause] divided by c equals a divided by c plus b divided by c"

DT1_4_e: The process of solving equations can sometimes be simplified by factoring out common factors: $0=3x^3-6x^2$; $0=3x^2(x-2)$; $x_1=0$, $x_2=2$

DT1_5_e: In this case, we use the "Zero-Product Property [:]
AB=0 if and only if A=0 or B=0"(Stewart, Redlin & Watson 2002: 48).

DT1_1_d: Eine Summe wird mit einer Zahl multipliziert, indem man jeden Summanden mit dieser Zahl multipliziert und die Ergebnisse addiert: a·(b+c)=a·b+a·c. Dies wird auch „Auflösen der Klammer" genannt.
Lies: „a mal Klammer auf b plus c Klammer zu gleich a mal b plus a mal c)

DT1_2_d: Der umgekehrte Vorgang heißt „Ausklammern von a": a·b+a·c= a·(b+c)

DT1_3_d: Eine Summe wird durch eine Zahl dividiert, indem man jeden Summanden durch diese Zahl dividiert und die Ergebnisse addiert:
$\frac{a+b}{c}=\frac{a}{c}+\frac{b}{c}$. E.g., $\frac{104}{4}=\frac{100+4}{4}=\frac{100}{4}+\frac{4}{4}=25+1=26$
Lies: "a plus b [Pause] dividiert durch c gleich a durch c plus b durch c"

DT1_4_d: Das Lösen von Gleichungen kann manchmal vereinfacht werden, indem man gemeinsame Faktoren ausklammert: $0=3x^3-6x^2$; $0=3x^2(x-2)$; $x_1=0$, $x_2=2$

DT1_5_d: In diesem Fall benutzen wir den „Nullproduktsatz [...]: Das Produkt zweier Zahlen ist genau dann null, wenn mindestens einer der Faktoren null ist" [...]
$a \cdot b = 0 \Leftrightarrow a = 0 \vee b = 0$" Kreul & Ziebarth 2016: 152)

[The German wording on this page is influenced by a worksheet distributed by my mathematics teacher H.-J. K. in 1973. The wording used in a German bestseller sold more than 1 million times (Kreul & Ziebarth 2016: 2) is very similar.]

PC1: Percentages 1 / Prozentrechnung 1

PC1_1_e: "One percent (1%) of a quantity G is one hundredth of the quantity G" (Kreul & Ziebarth 2026: 339, translated by SK).

PC1_2_e: "A percentage is a convenient way of writing fractions that have a denominator of 100. [...] eg $7\% = \frac{7}{100}$" (McSeveny, Conway, Wilkes & Smith 2014: 175)

PC1_3_e: To convert a percentage to a fraction or a decimal, "first write it as a fraction with denominator 100" (McSeveny, Conway, Wilkes & Smith 2014: 175):

$75\% = \frac{75}{100} = 0.75 \qquad 75\% = \frac{75}{100} = \frac{3 \cdot 25}{4 \cdot 25} = \frac{3}{4}$

PC1_4_e: "To find a percentage of a quantity, write the percentage as a decimal (or fraction) and multiply by the quantity" (McSeveny, Conway, Wilkes & Smith 2014: 180): e.g., 12% of 200kg=0.12×200kg=24kg

PC1_5_e: "To write one quantity as a percentage of another: [...] Make sure that the two quantities are in the same units. [...] Write the first as a fraction of the second. [...] Change the fraction to a percentage" (McSeveny, Conway, Wilkes & Smith 2014: 181): What percentage of 2.5kg are 20g? 20g of 2.5kg are 0.020kg of 2.5kg , so that we obtain $\frac{0.020kg}{2.5kg} = \frac{0.020 \cdot 4}{2.5 \cdot 4} = \frac{0.080}{10} = \frac{0.080 \cdot 10}{10 \cdot 10} = \frac{0.80}{100} = 0.80\%$

PC1_6_e: "To say `b is p percent of a´ is to say $b = \left(\frac{p}{100}\right)a$" (Esty 1997: 276).

PC1_1_d: „Ein Prozent (1%) einer Größe G ist der hundertste Teil von G" (Kreul & Ziebarth 2026: 339).

PC1_2_d: „Ein Prozentsatz ist eine angenehme Art, Brüche zu schreiben, die den Nenner 100 haben [...]" (McSeveny, Conway, Wilkes & Smith 2014: 175, übersetzt von SK). [Beispiel siehe oben]

PC1_3_d: „Um einen Prozentsatz in einen gemeinen Bruch oder einen Dezimalbruch umzuwandeln, „schreibe ihn zuerst als gemeinen Bruch mit dem Nenner 100" (McSeveny, Conway, Wilkes & Smith 2014: 175, übersetzt von SK). [Bsp. siehe oben]

PC1_4_d: „Um den Prozentwert einer Größe zu finden, schreibe den Prozentsatz als Dezimalzahl (oder als gemeinen Bruch) und mutipliziere mit der Größe" (McSeveny, Conway, Wilkes & Smith 2014: 180, übersetzt von SK):
z.B. 12% von 200kg=0,12·200kg=24kg

PC1_5_d: „Um eine Größe als Prozentsatz einer anderen zu schreiben: [...] Stelle sicher, dass die zwei Größen in denselben Einheiten gegeben sind. [...] Schreibe die erste Größe als Bruchteil der zweiten. [...] Wandle den Bruch in einen Prozentsatz um" (McSeveny, Conway, Wilkes & Smith 2014: 181, übersetzt von SK):
Welcher Prozentsatz von 2.5kg sind 20g? 20g von 2,5kg sind 0,020kg von 2,5kg, also

$\frac{0{,}020kg}{2{,}5kg} = \frac{0{,}020 \cdot 4}{2{,}5 \cdot 4} = \frac{0{,}080}{10} = \frac{0{,}080 \cdot 10}{10 \cdot 10} = \frac{0{,}80}{100} = 0{,}80\%$

PC1_6_d: "`b ist p Prozent von a´ heißt $b = \left(\frac{p}{100}\right)a$" (Esty 1997: 276, Übers.: SK).

PC2: Percentages 2 / Prozentrechnung 2

PC2_1_e: „To compare *b* to the standard *a*, the percent *p* is given by $\frac{b}{a}=\frac{p}{100}$. [...] The standard is in the denominator." (Esty 1997: 277, boldface removed).

PS2_2_e: "To \`increase´, \`gain´ or \`go up´ p percent means the final amount is the standard multiplied by a factor of 1 + p/100" (Esty 1997: 278).

PS2_3_e: "To \`decrease´, \`lose´, or \`go down´ p percent means the final amount is the standard multiplied by a factor of 1 - p/100" (Esty 1997: 278).

PS2_4_e: "\`*B* is 30% more than *A*´ [...] *B* is being compared to *A*, so *A* is the standard. *A* is therefore regarded as 100%, so *B* is 130% of *A*: *B*=1.3*A*" (Esty 1997: 278, italics in the source).

PS2_5_e: "The value *P* of the percentage of a standard [*G*] is a part of the standard. The percentage *p* states how many hundredths of the standard [*G*] correspond to the value *P*: [...] $\frac{P}{G}=\frac{p}{100}$ (Kreul & Ziebarth 2016: 340, italics in the source, translated by SK). [There is no single English word for the German term "Prozentwert", just the expression "percentage of a quantity" (McSeveny, Conway, Wilkes & Smith 2014: 180).]

PS2_6_e: "Calculations involving percentages are a special case of doing fractions with the denominator of 100" (Kreul & Ziebarth 2016: 340, übersetzt von SK).

PC2_1_d: „Um *b* mit dem Standard [oder Grundwert] *a* zu vergleichen, ist der Prozentsatz *p* gegeben durch $\frac{b}{a}=\frac{p}{100}$. [...] Der Standard [Grundwert] steht im Nenner" (Esty 1997: 277, Fettdruck entfernt, übersetzt von SK). [Kreul und Ziebarth bezeichnen „den Grundwert als Bezugspunkt“ (Kreul & Ziebarth 2016: 344), was „Standard“ bedeutet.]

PS2_2_d: „\`zunehmen´, \`gewinnen´ oder \`ansteigen´ um p Prozent bedeutet, dass der Endbetrag gleich dem Standard mal dem Faktor 1 + p/100 ist" (Esty 1997: 278).

PS2_3_d: „\`abnehmen´, \`verlieren´ oder \`absteigen´ um p Prozent bedeutet, dass der Endbetrag gleich dem Standard mal dem Faktor 1 - p/100 ist" (Esty 1997: 278).

PS2_4_d: "\`*B* ist 30% mehr als *A*´ [...] *B* wird verglichen mit *A*, so dass *A* der Standard [oder Grundwert] ist. *A* wird daher als 100% betrachtet, so dass *B* 130% von *A* ist: *B*=1,3*A*" (Esty 1997: 278, kursiv im Original, überstzt von SK).

PS2_5_d: „Der Prozentwert *P* ist ein Teil des Grundwertes [*G*]. Der Prozentsatz *p* gibt an, wie viel Hundertstel des Grundwertes dem Prozentwert entsprechen: [...] $\frac{P}{G}=\frac{p}{100}$ (Kreul & Ziebarth 2016: 340, kursiv im Original).

PS2_6_d: „Die Prozentrechnung ist ein Spezialfall der Bruchrechnung mit dem Nenner 100“ (Kreul & Ziebarth 2016: 340).

PW1: Powers 1 / Potenzen 1

PW1_1_e: “If *a* is any real number and *n* is a positive integer, then the ***n*th power** of *a* is $a^n = a \cdot a \cdot \ldots \cdot a$ [*n* factors] [...].
The number *a* is called the **base** and *n* is called the **exponent**“ (Stewart, Redlin & Watson 2002: 12, 15, italics and boldface in the source).

PW1_2_c: Instead of “*n*th power of *a*”, other manners of speaking are also common: “Read `$e^x$´ as `e to the x´” (Esty 1997: 329). “m^k“ can also be read as “*m* to the power *k*” (Beck & Geoghegan 2010: 175, italics in the source).

PW1_3_d: $a^0 := 1$: *a* to the power of 0 is 1 by definition. The sign „:=” is read :=“is defined by”(Beck & Geoghegan 2010: 175).

PW1_4_e: $a^1 := a$: *a* to the power of 1 is *a* by definition (is defined by *a*).

PW1_5_e: “To multiply powers of the same number, add the exponents“ (Stewart, Redlin & Watson 2002: 15): $a^n \cdot a^m = a^{n+m}$.

PW1_6_e: “a to the power of n times a to the power of m equals a to the power of [pause] n plus m”

PW1_7_e: “To divide two powers of the same number, subtract the exponents” (Stewart, Redlin & Watson 2002: 15): $\frac{a^n}{a^m} = a^{n-m}$

PW1_8_e: “a to the power of n divided by a to the power of m equals a to the power of [pause] n minus m”

PW1_1_d: “Wenn *a* eine beliebige reelle Zahl ist und *n* eine positive ganze Zahl, dann ist die ***n*-te Potenz** von *a* gleich $a^n = a \cdot a \cdot \ldots \cdot a$ [*n* Faktoren] [...].
Die Zahl *a* wird die **Basis** genannt und *n* der **Exponent**“ (Stewart, Redlin & Watson 2002: 12, 15, italics and boldface in the source, übersetzt von SK).

PW1_2_d: Statt „*n*-te Potenz von *a*“ sind auch andere Sprechweisen gebräuchlich: “Lies `$e^x$´ als `e hoch x´” (Esty 1997: 329, übersetzt von SK). “m^k“ kann auch gelesen werden als “*m* zur *k*-ten Potenz ” (Beck & Geoghegan 2010: 175, kursiv im Original, übersetzt von SK).

PW1_3_d: $a^0 := 1$: *a* hoch null ist definitionsgemäß 1. Das Zeichen „:=” wird gelesen als “wird definiert als” (Kreul & Ziebarth 2016: 194).

PW1_4_d: $a^1 := a$: *a* hoch 1 ist definitionsgemäß *a* (wird als *a* definiert).

PW1_5_d: Potenzen mit gleicher Basis werden multipliziert, indem man die Basis beibehält und die Exponenten addiert: $a^n \cdot a^m = a^{n+m}$

PW1_6_d: „a hoch n mal a hoch m gleich a hoch [Pause] n plus m“

PW1_7_d: „Potenzen mit gleicher Basis werden dividiert, indem man die Basis beibehält und die Exponenten subtrahiert“ (Kreul & Ziebarth 2016: 198): $\frac{a^n}{a^m} = a^{n-m}$

PW1_8_d: „a hoch n geteilt durch a hoch m gleich a hoch [Pause] n minus m“

PW2: Powers 2 / Potenzen 2

PW2_1_e: “To raise a power to a new power, multiply the exponents” (Stewart, Redlin & Watson 2002: 15): $\left(a^n\right)^m = a^{n \cdot m}$

PW2_2_e: “a to the power of n [pause] to the power of m equals a to the power of [pause] n times m”

PW2_3_e: “To raise a product to a power, raise each factor to the power” (Stewart, Redlin & Watson 2002: 15): $(a \cdot b)^n = a^n \cdot b^n$

PW2_4_e: “a times b [pause] to the power of n equals a to the power of n [pause] times b to the power of n”

PW2_5_e: “To raise a quotient to a power, raise both numerator and denominator to the power” (Stewart, Redlin & Watson 2002: 15): $\left(\frac{a}{b}\right)^n = \frac{a^n}{b^n}$

PW2_6_e: „a over b [pause] to the power of n equals a to the power of n [pause] over [pause] b to the power of n“

PW2_1_d: „Eine Potenz wird nochmals potenziert, indem man die Basis beibehält und die Exponenten multipliziert“ (Kreul & Ziebarth 2016: 200): $\left(a^n\right)^m = a^{n \cdot m}$

PW2_2_d: „a hoch n [Pause] hoch m gleich a hoch [Pause] n mal m“

PW2_3_d: Ein Produkt wird potenziert, indem man die Faktoren einzeln potenziert und die Ergebnisse multipliziert: $(a \cdot b)^n = a^n \cdot b^n$

PW2_4_d: „a mal b [Pause] hoch n gleich a hoch n [Pause] mal b hoch n“

PW2_5_d: „Potenzen mit gleichen Exponenten werden dividiert, indem man die Basen durcheinander teilt und den gemeinsamen Exponenten beibehält“ (Kreul & Ziebarth 2016: 200): $\frac{a^n}{b^n} = \left(\frac{a}{b}\right)^n$

PW2_6_d: “a hoch n [Pause] durch [Pause] b hoch n gleich a durch b [Pause] hoch n ”

PW3: Powers 3 / Potenzen 3

PW3_1_e: "One over [pause] a to the power of n equals a to the power of negative n."

PW3_1_d: "Eins durch [Pause] a hoch n gleich a hoch minus n." $\frac{1}{a^n} = a^{-n}$

PW3_2_e: The reciprocal of a power equals the power raised to the negative exponent.

PW3_2_d: „Beim Bilden des Kehrwertes einer Potenz ändert sich das Vorzeichen des Exponenten“ (Kreul & Ziebarth 2016: 204).

PW3_3_e: "To raise a fraction to a negative power, invert the fraction and change the sign of the exponent“(Stewart, Redlin & Watson 2002: 17).

PW3_3_d: "Statt einen Bruch mit einem neativen Exponenten zu potenzieren, kann man auch den Kehrwert des Bruches mit dem entpsrechenden possitiven Exponenten potenzieren" (Kreul & Ziebarth 2016: 206). $\left(\frac{a}{b}\right)^{-n} = \left(\frac{b}{a}\right)^{n}$

PW3_4_e: „"a over b [pause] to the power of negative n equals b over a [pause] to the power of n."

PW3_4_d: „a durch b [Pause] hoch minus n gleich b durch a [Pause] hoch n."

PW3_4_e: "The *n*th root of *x* is the number that, when raised to the *n*th power, gives *x*“ (Stewart, Redlin & Watson 2002: 17, italics in the source).

PW3_4_d: „Die *n*-te Wurzel von *x* ist die Zahl, die mit *n* potenziert *x* ergibt“ (Stewart, Redlin & Watson 2002: 17, kursiv im Original, überstzt von SK).

$$\left(\sqrt[n]{x}\right)^n = x$$

PW3_5_e: "The nth root of x equals x to the power of [pause] 1 over *n*."

PW3_5_d: „Die *n*-te Wurzel von *x* ist gleich *x* hoch [Pause] 1 durch *n*.“

$$\sqrt[n]{x} = x^{\frac{1}{n}}$$

MF1 Compound Interest 1 / Compound Interest 1

MF1_1_e: "Let an amount *P* (for principal) be invested at an interest rate of *i* per time period and compounded for *n* time periods [...]. At the end of n time periods its amount, *A*, satisfies $A = P(1+i)^n$" (Esty 1997: 281, italics in the source).

MF1_2_e: "If a starting capital K_0 is left untouched on an account earning *p* % [annual] interest for *n* years, it increases to the amount $K_n = K_0 \cdot (1+\frac{p}{100})^n$ (Kreul & Ziebarth 2016: 351, italics in the source, translated by SK).

MF1_1_d: „Man lasse einen Betrag *P* (für \`principal´ [=Kapitalbetrag]) investiert zu einem Zinssatz von *i* und schlage *n* Zeitperioden lang die Zinsen dem Kapital zu [...]. Am Ende von *n* Zeitperioden erfüllt der Endbetrag, *A*, die Gleichung $A = P(1+i)^n$" (Esty 1997: 281, kursiv im Original, übersetzt von SK).

MF1_2_d: „Verbleibt ein Anfangskapital K_0 bei einem [jährlichen] Zinssatz von *p* % für *n* Jahre unangetatstet auf einem Konto, so wächst es in dieser Zeit auf den Betrag $K_n = K_0 \cdot (1+\frac{p}{100})^n$ an" (Kreul & Ziebarth 2016: 351, kursiv im Original).

RD1: Rounding 1 / Runden 1

RD1_1_e: "Any [...][number] can be *rounded* off to a certain number of digits. In rounding off, it is common to look at the next to last digit. If it is followed by a digit from 0 to 4, it is often rounded down and so stays the same. If it is followed by a digit from 5 to 9, it is often rounded up and increased by one" (Herman 2020: 80, italics in the source).

RD1_2_e: "This method of rounding actually skews (or tilts) numbers higher on the average. To avoid that, when the last digit is 5, the next to last digit is rounded up when it is odd and down when it is even, which is known as bankers´ rounding, because bankers often use it. So, 3.5 rounds up to 4, 4.5 rounds down to 4, and 5.5 rounds up to 6 and so on, which explains why this method is also known as the round-to-even rule" (Herman 2020: 80).

RD1_3_e: "In these cases, it is usually suggested to round to `even´, i.e., use the estimates 0, 2, 4, 6, 8 for the numbers 0.5, 2.5, 4.5, 6.5, 8.5 and the estimates 2, 4, 6, 8, 10 for the numbers 1.5, 3.5, 5.5, 7.5 9.5. At the same time, the aim is achieved that the errors are sometimes positive, sometimes negative, so that some compensation is to be expected in the course of a longer, continuous calculation (Bruns 1903: 28, translated by SK).

RD1_4_d: If 1.2347 is rounded to three decimal places, it is rounded up to 1.235, because the digit on the last remaining decimal place is followed by a 7. If 1.2347 is rounded to two decimal places, it is rounded down to 1.23, because the digit on the last remaining decimal place is followed by a 4.

RD1_1_d: „Jede [...][Zahl] kann auf eine bestimmte Zahl von Dezimalstellen *gerundet* werden. Beim Runden betrachtet man gewöhnlich die vorletzte Stelle. Falls auf diese eine Ziffer von 0 bis 4 folgt, wird sie oft abgerundet und bleibt somit gleich. Falls auf diese eine Ziffer von 5 bis 9 folgt, wird sie oft aufgerundet und um eins erhöht" (Herman 2020: 80, kursiv im Original, übersetzt von SK).

RD1_2_d: „Diese Rundungsmethode verzerrt (oder neigt) tatsächlich Zahlen im Durchschnitt höher. Um das zu vermeiden, wenn die letzte Ziffer 5 ist, wird die vorletzte Ziffer aufgerundet, wenn sie ungerade ist, und abgerundet, wenn sie gerade ist, was als Bankiers-Rundung bekannt ist, weil diese die Methode oft nutzen. Daher rundet man 3,5 auf 4 auf, 4,5 auf 4 ab, und man rundet 5,5 auf 6 auf und so weiter, was erklärt, warum diese Methode auch Runden-auf-gerade-Zahl bekannt ist (Herman 2020: 80, übersetzt von SK).

RD1_3_d: „Gewöhnlich wird für diese Fälle vorgeschlagen, auf `gerade´ abzurunden, d.h. für die Beträge 0.5, 2.5, 4.5, 6.5, 8.5 die Zahlen 0, 2, 4, 6, 8 und für die Beträge 1.5, 3.5, 5.5, 7.5 9.5 die Zahlen 2, 4, 6, 8, 10 anzusetzen. Man erreicht dabei zugleich, daß die Fehler bald positiv, bald negativ ausfallen, daß also innerhalb einer längeren zusammenhängenden Rechnung im allgemeinen eine gewisse Ausgleichung zu erwarten ist" (Bruns 1903: 28).

RD1_4_d: Falls 1,2347 auf drei Dezimalstellen gerundet wird, wird auf 1,235 aufgerundet, weil auf die letzte verbleibende Stelle eine 7 folgt. Falls 1,2347 auf zwei Dezimalstellen gerundet wird, wird auf 1,23 abgerundet, weil auf die letzte verbleibende Stelle eine 4 folgt.

RD2: Rounding 2 / Runden 2

RD2_1_e: Rounding according to (the German Industrial Standard) DIN 1333: "The last, not required digits of a decimal fraction are omitted, if the first of the discarded digits is 0, 1, 2, 3, or 4: The fraction is *rounded down*. If the first of the discarded digits is 5, 6, 7, 8 or 9, the last digit to remain is increased by 1: The fraction is *rounded up*" (Kreul & Ziebarth 2016: 70, italics in the source).

RD2_2_e: "As a matter of principle, it is advisable to use the initial values in calculations [and proceed without rounding] as long as possible and only round off the final result to the desired number of digits" (Kreul & Ziebarth 2016: 72).

RD2_3_e: "[T]o make the final result as accurate as possible, *you should wait until the last step to round off your answer*" (Stewart, Redlin & Watson 2002: xxi, italics in the source).

RD2_4_e: Rounding allows to estimate results: $311/330 \times 480 \times 6.3$=?
"An approximate answer is readily obtained as follows:" (Burton 1994: 2)
"(nearly 1) × (just under 500) × (just over6)= slightly under 3000" (Burton 1994: 2)
"The 480 has been rounded up and 6.3 rounded down in a way that should roughly cancel out the resulting errors" (Burton 1994: 2). The exact result is 2849.890909. Indeed, this is slightly under 3000. Our approximate answer is only about 5.3% higher.

RD2_5_e: 2.4578 ist rounded to two decimal places by replacing the digit on the second decimal place, 5, by 6. 5.6432 is rounded to two decimal places by leaving the digit on the second decimal place, 4, unchanged.

RD2_1_d: Rundung nach DIN 1333: „Die letzten, nicht benötigten Ziffern eines Dezimalbruches werden weggelassen, wenn die erste der weg-gestrichenen Ziffern eine 0, 1, 2, 3, oder 4 ist: Der Bruch wird *abgerundet*. Ist die erste der weggestrichenen Ziffern eine 5, 6 , 7, 8 oder 9, so wird die letzte stehen bleibende Ziffer um 1 erhöht: Der Bruch wird *aufgerundet*" (Kreul & Ziebarth 2016: 70, kursiv im Original).

RD2_2_d: „Grundsätzlich sollte man so lange wie möglich mit den Ausgangszahlen [ohne Rundung] weiterrechnen und nur das Endergebnis auf die nötige Stellenzahl runden" (Kreul & Ziebarth 2016: 72).

RD_3_d: „Um das Endergebnis so genau wie möglich zu machen, *sollten Sie bis zum letzten Schritt damit warten, Ihre Antwort zu runden*" (Stewart, Redlin & Watson 2002: xxi, kursiv im original, übersetzt durch SK).

RD2_4_d: Mit Runden kann man Resultate schätzen: $311/330 \times 480 \times 6{,}3$=?
„Eine angenäherte Lösung erhält man ohne weiteres wie folgt:" (Burton 1994: 2)
„(fast 1) × (gerade eben unter 500) × (gerade eben über 6) = etwas unter 3000" (Burton 1994: 2) „Die 480 ist aufgerundet worden und 6,3 abgerundet worden auf eine Art, die ganz grob die resultierenden Fehler aufheben sollte" (Burton 1994: 2). Die genaue Lösung ist 2849.890909. Das ist tatsächlich etwas unter 3000. Unser Näherungswert ist nur ungefähr 5,3% höher.

RD2_5_d: Die Zahl 2,4578wird auf zwei Dezimalstellen gerundet, indem man die Ziffer auf der zweiten Dezimalstelle, 5, durch eine 6 ersetzt. Die Zahl 5,6432 wird auf zwei Dezimalstellen gerundet, indem man die Ziffer auf der zweiten Dezimalstelle, 4, unverändert lässt.

LG1 Logarithms 1 / Logarithmen 1

LG1_1_e: In the equation $a^m = q$ [“a to the power of m equals q”] the exponent m is unknown while the positive numbers a and q are known. „This **exponent we are looking for** gets a name. It is called the **logarithm of q with respect to the base a**“ (Läuger 1992: 69, boldface in the source, translated by SK). $m = \log_a(q)$

LG1_2_e: In the equation $10^m=1000$ (“ten to the power of m equals 1000”) the exponent m is called the logarithm of 1000 with respect to the base 10. $m=\log_{10}(1000)$. m=3

LG1_3_e: In this example, the logarithm m is the exponent required to write 1000 as a power of 10.

LG1_4_e: “The log in base *e*, $\log_e$, [$e \approx 2.718$] is called the *natural logarithm* and has a special symbol `ln´ (`l´ `n´). It is colloquially called the `natural log´ or just log (and then pronounced `log´) when the context is clear” (Herman 2020: 71, italics in the source).

LG1_5_e: “The *logarithm, A*, of a number, *N*, to the base *a* is the exponent of the power of *a* that is equal to *N*. we write $A=\log_a N$. Consequently, the equations $a^A=N$ and $A=\log_a N$ are equivalent” (Bronshtein & Semendyayev 1973: 156, italics in the source).

LG1_6_e: “The logarithm of a number *b* with respect to the base *a* is the exponent *c*, to which the base *a* has to be raised to get the numerus [the number whose logarithm has to be calculated] *b*” (Kreul & Ziebarth 2016: 260, italics in the source, transl. by SK). “$a^c = b \Leftrightarrow c = \log_a(b), 5^3 = 125 \Leftrightarrow 3 = \log_5(125)$“ (Kreul & Ziebarth 2016: 260, 261)

LG1_1_d: In der Gleichung $a^m = q$ [„a hoch m gleich q“] mit bekannten, positiven Zahlen a und q sucht man den Exponenten m. „Diesem **gesuchten Exponenten** m gibt man jetzt einen Namen, man nennt ihn **Logarithmus von q zur Basis a**“ (Läuger 1992: 69, Fettdruck im Original). $m = \log_a(q)$

LG1_2_d: In der Gleichung $10^m=1000$ („Zehn hoch m gleich Tausend”), wird der Exponent m Logarithmus von 1000 zur Basis 10 genannt. $m=\log_{10}(1000)$. m=3

LG1_3_d: In diesem Beispiel ist der Logarithmus m der erforderliche Exponent, um 1000 als Potenz mit der Basis 10 darzustellen.

LG1_4_d: „Der Logarithmus zur Basis *e*, $\log_e$, [$e \approx 2.718$] wird *natürlicher Logarithmus* genannt und hat ein besonderes Symbol `ln´ (`l´ `n´). Umgangsprachlich wird er `natürlicher log´ oder nur log genannt, wenn der Zusammenhang klar ist“ (Herman 2020: 71, Kursivschrift im Original, übersetzt von SK).

LG1_5_d: „Der *Logarithmus, A*, einer Zahl, *N*, zur Basis *a* ist der Exponent der Potenz von *a*, die gleich *N* ist. we write $A=\log_a N$. Consequently, the equations $a^A=N$ and $A=\log_a N$ are equivalent” (Bronshtein & Semendyayev 1973: 156, kursiv im Original, übersetzt von SK).

LG1_6_d: „Der Logarithmus einer Zahl *b* zur Basis *a* ist derjenige Exponent *c*, mit dem die Basis *a* potenziert werden muss, um den Numerus [Zahl, von der ein Logarithmus bestimmt werden soll] *b* zu erhalten“ (Kreul & Ziebarth 2016: 260, kursiv im Original). „$a^c = b \Leftrightarrow c = \log_a(b), 5^3 = 125 \Leftrightarrow 3 = \log_5(125)$“ (Kreul & Ziebarth 2016: 260, 261)

LG2 Logarithms 2 / Logarithmen 2

LG2_1_e: "$\log_a x = y \Leftrightarrow a^y = x$ [...] $\log_a x$ is the exponent to which the base *a* must be raised to give *x*" (Stewart, Redlin & Watson 2002: 349, italics in the source).

LG2_2_e: "It's important to understand that $\log_a x$ is an *Exponent* [...] $\log_2 32=5$ because $2^5=32$" (Stewart, Redlin & Watson 2002: 349, italics in the source).

LG2_3_e: „The logarithm of a product of numbers is the sum of the logarithms of the numbers" (Stewart, Redlin & Watson 2002: 359): $\log_a(x \cdot y) = \log_a(x) + \log_a(y)$.

LG2_4_e: "The logarithm of a quotient of numbers is the difference of the logarithms of the numbers" (Stewart, Redlin & Watson 2002: 359): $\log_a(x/y) = \log_a(x) - \log_a(y)$.

LG2_5_e: "The logarithm of a power of a number is the exponent times the logarithm of the number" (Stewart, Redlin & Watson 2002: 359): $\log_a(x^c) = c \cdot \log_a(x)$.

LG2_6_e: The logarithm of the base *a* with respect to the base *a* is 1: $\log_a(a) = 1$. The logarithm of 1 with respect to any permissible base is 0.; $\log_a(1) = 0$

LG2_7_e: "$\log_a a^x = x$ We must raise *a* to the power *x* to get a^x " (Stewart, Redlin & Watson 2002: 353, italics in the source).

LG2_8_e: "Logarithms are exponents. When $10^x=y$, the log of the number y is the exponent x" (Esty1997: 329)

$a>0$, $a\neq 1$, $x>0$, $y>0$, c can be any real number / c kann jede reelle Zahl sein

----

LG2_1_d: „$\log_a x = y \Leftrightarrow a^y = x$ [...] $\log_a x$ ist der Exponent, mit dem die Basis *a* zu potenzieren ist, um *x* zu ergeben" (Stewart, Redlin & Watson 2002: 349, übersetzt durch SK).

LG2_2_d: „Es ist wichtig zu vestehen, dass $\log_a x$ ein *exponent* ist [...] $\log_2 32=5$ weil $2^5=32$" (Stewart, Redlin & Watson 2002: 349, italics in the source, übersetzt durch SK).

LG2_3_d: „Der Logarithmus eines Produktes von Zahlen ist die Summe der Logarithmen der einzelnen Zahlen" (Stewart, Redlin & Watson 2002: 359, übersetzt durch SK): $\log_a(x \cdot y) = \log_a(x) + \log_a(y)$.

LG2_4_d: „Der Logaritmus eines Quotienten von Zahlen ist die Differenz der Logarithmen der einzelnen Zahlen" (Stewart, Redlin & Watson 2002: 359, übersetzt durch SK): $\log_a(x/y) = \log_a(x) - \log_a(y)$.

LG2_5_d: „Der Logarithmus einer Potenz einer Zahl ist der Exponent multipliziert mit dem Logarithmus der Zahl" (Stewart, Redlin & Watson 2002: 359, übersetzt durch SK): $\log_a(x^c) = c \cdot \log_a(x)$.

LG2_6_d: Der Logarithmus der Basis a zur Basis a ist 1: $\log_a(a) = 1$. Der Logarithmus von 1 zu jeder erlaubten Basis *a* ist 0: $\log_a(1) = 0$.

LG2_7_d: „$\log_a a^x = x$ Wir müssen *a* mit *x* potenzieren, um a^x *zu erhalten*" (Stewart, Redlin & Watson 2002: 353, kursiv im Original, übersetzt von SK).

LG2_8_d: „Logarithmen sind Exponenten. When $10^x=y$, dann ist der log der Zahl y gleich dem Exponenten x (Esty 1997: 329, übestzt von SK).

SN1: Scientific Notation 1 / Wissenschaftliche Schreibweise von Zahlen 1

SN1_1_e: "Scientific notation is a way to write numbers that takes advantage of the powers of ten (Adams 2003: 5). "The idea is to rewrite a number as a number between 1 and 10, times 10 to the appropriate power"(Esty 2012: 293).

SN1_2_e: "Writing the length $1.1\mu m$ as $1.1\cdot 10^{-6}m$ is to express the value in scientific notation" (Kirkup 1994: 14, emphasis removed).

SN1_3_e: "Powers-of-10 Notation" (Romer 1976: 31)

SN1_4_e: Example: Munich has 1.5 million inhabitants, each of whom needs 120L of water per day (my estimate). How many cubic metres are that per day? Using 10^3L=1m^3 we obtain $1.5\times10^6\times1.2\times10^2$L=$1.5\times1.2\times10^{6+2}$L=$1.8\times10^8$L=$1.8\times10^5\cdot10^3$L=$1.8\times10^5$m^3.

SN1_5_e: "Multiplication or division of very large or very small numbers is most conveniently handled by first writing all numbers in powers-of-10 form" (Romer 1976: 31).

SN1_6_e: Example: $0.00045\times2100000=4.5\times10^{-4}\times2.1\times10^6=9.45\times10^{-4+6}=9.45\times10^2=945$

SN1_7_e: "A positive number *x* is said to be written in scientific notation if it is expressed as follows: $x=a\times10^n$ where $1\leq a<10$ and *n* is an integer"(Stewart, Redlin & Watson 2002: 18, italics in the source).

SN1_8_e: "7.39842×10^6 is more comprehensible and legible than seventythree billion nine hundred and eighty-four million and two hundred thousand" (Paulos 2001: 11).

SN1_1_d: "Wissenschaftliche Notation ist eine Art, Zahlen zu schreiben, die Zehnerpotenzen nutzt"(Adams 2003: 3, übersetzt von SK). „Die Idee ist es, eine Zahl umzuschreiben als eine Zahl zwischen 1 und 10, multipliziert mit der geeigneten Potenz von 10"(Esty 2012: 293, übersetzt von SK).

SN1_2_d: "Die Länge $1.1\mu m$ als $1.1\cdot 10^{-6}m$ zu schreiben, bedeutet den Wert in wissenschaftlicher Schreibweise auszudrücken" (Kirkup 1994: 14, emphasis removed, translated by SK).

SN1_3_d: "Zehnerpotenzenschreibweise"(Romer 1976: 31, übersetzt von SK)

SN1_4_d: Beispiel: München hat 1,5 Millionen Einwohner, von denen jeder 120L Wasser pro Tag braucht (meine Schätzung). Wie viele Kubikmeter sind das pro Tag? Wir erhalten $1{,}5\cdot10^6\cdot1{,}2\cdot10^2$L=$1{,}5\cdot1{,}2\cdot10^{6+2}$L=$1{,}8\cdot10^8$L=$1{,}8\cdot10^5\cdot10^3$L=$1{,}8\cdot10^5$m^3 unter Verwendung von 10^3L=1m^3.

SN1_5_d: „Die Multiplikation oder Division sehr großer oder sehr kleiner Zahlen wird am bequemsten gehandhabt, indem man zunächst alle Zahlen in Zehnerpotenzenschreibweise darstellt"(Romer 1976: 31, übersetzt von SK).

SN1_6_d: Beispiel: $0{,}00045\cdot2100000=4{,}5\cdot10^{-4}\cdot2{,}1\cdot10^6=9{,}45\cdot10^{-4+6}=9{,}45\cdot10^2=945$

SN1_7_d: "Man sagt von einer positiven Zahl, sie sei in wissenschaftlicher Notation geschrieben, wenn sie wie folgt ausgedrückt wird: $x=a\cdot10^n$ wobei $1\leq a<10$ und *n* eine ganze Zahl ist"(Stewart, Redlin & Watson 2002: 18, italics in the source, übers. v. SK).

SN1_8_d: „$7{,}39842\cdot10^{10}$ ist verständlicher und lesbarer als dreiundsiebzig Milliarden neunhundertvierundachtzig Millionen zweihunderttausend" (Paulos 1990: 52)

SN2: Scientific Notation 2 / Wissenschaftliche Schreibweise von Zahlen 2

SN2_1_e: "We write 257 as 2.57×10^2 [...]. The *exponent* tells us how many zeros are in the power of ten [...] and the *coefficient* multiplies the power of ten. This system is called *scientific notation*" (Weinstein & Adams 2008: 12, italics in the source). Example: $50000 = 5\times10000 = 5\times10^4$.

SN2_2_e: "The rules for multiplying and dividing with scientific notation are straightforward. When we multiply numbers, we multiply coefficients and add exponents. [...] When we divide numbers, we divide coefficients and we subtract exponents" (Weinstein & Adams 2008: 13). Examples: $2.5\times10^2\times3\times10^4=(2.5\times3)\times10^{2+4}=7.5\times10^6$; $(8\times10^6)\div(2\times10^4)=(8\div2)\times10^{6-4}=4\times10^2$

SN2_3_e: A city has 1.5 million inhabitants. Each of them drinks a cup of coffee per day, on the average, including two sugar cubes of 3.0g each. How many tons of sugar does this require per day? Using 10^3g=1000g=1kg and 10^3kg=1000kg=1t, we get in scientific notation:

$$1.5\times10^6\times2\times3.0g=1.5\times2\times10^3\times10^3\times3.0g=3\times3\times10^3\times10^3\,g=9.0\times10^3\,kg=9.0t$$

SN2_1_d: „Wir schreiben 257 als $2{,}57\cdot10^2$ [...]. Der *Exponent* sagt uns wie viele Nullen in der Zehnerpotenz auftreten [...] und der *Koeffizient* multipliziert die Zehnerpotenz. Dieses System wird *wissenschaftliche Notation* genannt" (Weinstein & Adams 2008: 12, kursiv im Original, übersetzt von SK). Beispiel: $50000 = 5\cdot10000 = 5\cdot10^4$.

SN2_2_d: „Die Regeln für das Multiplizieren und Dividieren mit der wissenschaftlichen Notation sind klar. Wenn wir Zahlen multiplizieren, multiplizieren wir Koeffizienten und addieren Exponenten. [...] Wenn wir Zahlen dividieren, dividieren wir Koeffizienten und subtrahieren Exponenten" (Weinstein & Adams 2008: 13, übersetzt von SK). Beispiele: $2{,}5\cdot10^2\cdot3\cdot10^4=(2{,}5\cdot3)\cdot10^{2+4}=7{,}5\cdot10^6$; $(8\cdot10^6)\div(2\cdot10^4)=(8\div2)\cdot10^{6-4}=4\cdot10^2$

SN2_3_d: Eine Großstadt hat 1,5 Millionen Einwohner. Jeder von diesen trinkt im Durchschnitt eine Tasse Kaffee pro Tag, mit jeweils zwei Zuckerwürfeln. Wie viele Tonnen Zucker werden dafür täglich benötigt? Verwenden wir 10^3g=1000g=1kg und 10^3kg=1000kg=1t, erhalten wir in wissenschaftlicher Notation:

$$1{,}5\cdot10^6\cdot2\cdot3{,}0g=1{,}5\cdot2\cdot10^3\cdot10^3\cdot3{,}0g=3{,}0\cdot3{,}0\cdot10^3\cdot10^3\,g=9{,}0\cdot10^3\,kg=9{,}0t$$

QM1: Quantities, Units, Measurements / Größen, Einheiten, Messungen

QM1_1_e: „The procedure of measuring a physical quantity consists in the determination of the number which states how often an agreed-upon unit is contained in the quantity. This ratio is called maesure or numerical value" (Bormann 1978: 6-7; translated by SK).

QM1_2_e: „Measuring some physical quantity means that it is numerically compared with a unit of measurement. This way, you obtain the product of numerical value times unit, e.g., three metre, six seconds etc." (Dorfmüller e.a. 1998: 27; emphasis removed, translated by SK).

QM1_3_e: "A given physical quantity A, such as length, time, or energy, is the product of a numerical value or measure {A} and a unit [A]. […] A={A}[A]" (Weil 2007: 70).

QM1_4_e: Definitions (Mende & Simon 2016: 11; translated by SK):
A: „value of the physical quantity"
{A}: „numerical value [or measure]", A is written in curly brackets.
[A]: „unit of the physical quantity", A is written in square brackets.

QM1_5_e: „Every value of a physical quantity is the product of a numerical value [or measure] and a unit" (Mende & Simon 2016: 11, translated by SK). A={A}[A]

QM1_6_e: Take A=24kg as an example. A is a physical quantity obtained, e.g., by measurement or by calculation with the help of a theory. {A}=24 is the numerical value or measure. [A]=kg is the unit. The multiplication sign between "24" and "kg" is omitted by tacit convention. I also omit the usual blank space between them.

QM1_1_d: „Der Vorgang der Messung einer physikalischen Größe besteht darin, die Zahl festzustellen, die angibt, wie oft die festgelegte Einheit in der Größe enthalten ist. Diese Verhältniszahl wird Maßzahl oder Zahlenwert genannt" (Bormann 1978: 6-7, Hervorhebung entfernt).

QM1_2_d: "Die Messung irgendeiner physikalischen Größe bedeutet, daß sie zahlenmäßig mit einer Maßeinheit verglichen wird. Man erhält so das Produkt Zahlenwert mal Maßeinheit, z.B. drei Meter, sechs Sekunden usw" (Dorfmüller e.a. 1998: 27).

QM1_3_d: "Eine gegebene physikalische Größe, wie zum Beispiel Länge, Zeit oder Energie, ist das Produkt eines Zahlenwertes {A} und einer Einheit [A]. […] A={A}[A]"(Weil 2007: 70, übersetzt von SK).

QM1_4_d: Definitionen (Mende & Simon 2016: 11): A: „Wert der physikalischen Größe"; {A}: „Zahlenwert [oder Maßzahl] der physikalischen Größe", A ist in geschweiften Klammern geschrieben. [A]: „Einheit der physikalischen Größe", A ist in eckigen Klammern geschrieben.

QM1_5_d: „Jeder Größenwert ist das Produkt aus Zahlenwert [oder Maßzahl] und Einheit" (Mende & Simon 2016: 11). A={A}[A]

QM1_6_d: Nehmen wir A=24kg als Beispiel. A ist eine physikalische Größe, die man z. B. als Wert einer Messung oder durch Brechnung mit Hilfe einer Theorie erhält. {A}=24 ist der Zahlenwert oder die Maßzahl. [A]=kg ist die Einheit. Das Multiplikationszeichen zwischen „24" und „kg" wird aufgrund stillschweigender Vereinbarung weggelassen. Ich lasse auch das übliche Leerzeichen zwischen diesen weg.

QM2: Quantities, Units, Measurements 2 / Größen, Einheiten, Messungen 2

QM2_1_e: According to the *Concise Oxford English Dictionary*, "measurement" (Stevenson & Waite 2011: 886) denotes both "the action of measuring" and "an amount, size, or extent as established by measuring". It is both process and result.

QM2_2_e: "Measurement is assignment of numbers to objects in such a way that these numbers correspond to the degree or amount of some quantity possessed by their objects" (Causey 1999: 528).

QM2_3_e: "We measure money in millions and time in hours" (Grimmett & Stirzaker 2007: 322) is an example of how mathematicians deal with units. Millions of Italian Lira or millions of US-Dollar?

QM2_1_d: Gemäß dem *Concise Oxford English Dictionary* bezeichnet das Wort "measurement" sowohl die "Handlung des Messens" als auch "eine Menge, Größe oder Ausdehnung, wie sie durch Messen festgestellt worden ist (Stevenson & Waite 2011: 886, übersetzt von SK). „Measurement" bezeichnet sowohl das Verfahren [Messen] als auch das Ergebnis [Messwert].

QM2_2_d: „Messen ist die Zuordnung von Zahlen zu Gegenständen in einer solchen Weise, dass die Zahlen dem Grad oder der Menge einer Größe entsprechen, welche die Gegenstände besitzen" (Causey 1999: 528, übersetzt von SK).

QM2_3_d: „Wir messen Geld in Millionen und Zeit in Stunden" (Grimmett & Stirzaker 2007: 322, übersetzt von SK) ist ein Beispiel dafür, wie Mathematiker mit Einheiten umgehen. Millionen italienischer Lire oder Millionen von US-Dollar?

CU1: Calculations Involving Units 1/ Rechnungen mit Einheiten 1

CU1_1_e: „Stating a measurement including a unit, e.g., stating the length 154 pm, means 154 times 1 picometer. Mathematically, this is a multiplication of a number by a unit" (Mortimer & Müller 2015: 28, translated by SK).

CU1_2_e: "If only numerical values are stated in a table of values [of a quantity], [...] the values [of the physical quantity] have been divided by the unit. Therefore, e.g., $\frac{Length}{pm}$ or Length/pm is given in the header of the table; the oblique stroke is to be understood as a fraction bar. Instead of the fraction bar, the word \`in´ can be used [...] (\`Length in pm´)"(Mortimer & Müller 2015: 28).

CU1_3_e: When you multiply or divide physical quantities, you can deal with the numerical values and the units separately. Just insert the multiplication signs between the numerical values and their assigned units and use the laws of multiplication and division (including power laws) as if the units were numbers. I have not yet found a book in which this is stated explicitly.

CU1_4_e: Calculate, for example, the mass m of $V=2.0cm^3$ of iron, which has a density of $\rho = 7860 kgm^{-3}$ (Mende & Simon 2016: 48). See below.

$$m = \rho \cdot V = 7860 kgm^{-3} \cdot 2cm^3 = 7860 \cdot kg \cdot m^{-3} \cdot 2 \cdot cm^3 = 7860 \cdot 2 \cdot kg \cdot \frac{cm^3}{m^3}$$

$$m = 7860 \cdot 2 \cdot kg \cdot \left(\frac{cm}{m}\right)^3 = 7860 \cdot 2 \cdot kg \cdot \left(\frac{1cm}{100cm}\right)^3 = 7860 \cdot 2 \cdot kg \cdot \frac{1}{1000 \cdot 1000} \approx 16g$$

CU1_1_d: „Die Angabe eines Messwertes mit Maßeinheit, zum Beispiel die Längenangabe 154 pm, bedeutet 154 mal 1 Picometer. Rechnerisch ist das die Multiplikation der Zahl mit der Maßeinheit" (Mortimer & Müller 2015: 28).

CU1_2_d: „Wenn in Tabellen [von Werten einer Größe] nur Zahlenwerte angegeben werden, [...] sind die Werte [der physikalischen Größen] durch die Maßeinheit dividiert worden. Deshalb steht dann im Tabellenkopf zum Beispiel $\frac{Länge}{pm}$ oder Länge/pm; der Schrägstrich ist als Bruchstrich zu verstehen. Statt des Bruchstrichs kann man auch das Wort \`in´ verwenden [...] (\`Länge in pm´)" (Mortimer & Müller 2015: 28).

CU1_3_d: Wenn man physikalische Größen multipliziert oder dividiert, kann man die Maßzahlen und die Einheiten getrennt bearbeiten. Man setze einfach Multiplikationszeichen zwischen die Maßzahlen und die zugeordneten Einheiten und benutze die Gesetze der Multiplikation und Division (einschließlich Potenzrechenregeln) als ob die Einheiten Zahlen wäre. Ich habe noch kein Buch gefunden, in welchem dies ausdrücklich dargelegt ist.

CU1_4_d: Man berechne zum Beispiel die Masse m von $V=2{,}0cm^3$ Eisen, welches eine Dichte von $\rho = 7860 kgm^{-3}$ (Mende & Simon 2016: 48) hat. Siehe oben.

CU2: Calculations Involving Units 1/ Rechnungen mit Einheiten 1

CU2_1_e: The rules for the manipulation of powers can be applied to units and their powers. $1m^3=1\times(100cm)^3=1\times100^3*cm^3=1\times(10^2)^3\times cm^3=1\times10^6\times cm^3=1000000cm^3$. One cubic metre contains one million cubic centimetres.

CU2_2_e: "[W]e must know how to convert quantities from one unit to another. [...] [W]e produce no change in a quantity if we simultaneously multiply and divide it by equal quantities, even though the quantities are expressed in different units. The [undesired] units [...] `cancel´ one another" [...] Another way of describing this process is this. The [original] quantity [...] has been multiplied by a ratio whose value is 1 [...] and multiplication by the number 1 produces no change. This is the true meaning of the term `*conversion factor*´, a quantity that is equal to 1 and that can be introduced as a multiplicative factor for the sole purpose of changing units" (Romer 1976: 36).

CU2_3_e: In the following example, devised by SK, we multiply the original quantity by two factors of 1, each suitably expressed as a ratio with the undesired unit in such a position that the undesired units of the original quantity are cancelled out.

$$36\frac{km}{h}=36\frac{km}{h}\cdot1\cdot1=36\frac{km}{h}\cdot\frac{h}{3600s}\cdot1=36\frac{km}{h}\cdot\frac{h}{3600s}\frac{1000m}{km}=36\frac{m}{s}\cdot\frac{1000}{3600}=10\frac{m}{s}$$

CU2_1_d: Auf Einheiten und ihre Potenzen können die Potenzrechenregeln angewandt werden. $1m^3=1\cdot(100cm)^3=1\cdot100^3*cm^3=1\cdot(10^2)^3\cdot cm^3=1\cdot10^6\cdot cm^3=1000000cm^3$. Ein Kubikmeter enthält eine Million Kubikzentimeter.

CU2_2_d: „Wir müssen nun wissen, wie wir Größen von einer Einheit auf eine andere umrechnen. [...] Wir erzeugen keine Änderung in einer Größe, wenn wir sie gleichzeitig mit derselben Größe multiplizieren und durch diese Größe dividieren, selbst wenn die Größen in unterschiedlichen Einheiten ausgedrückt sind. Die [unerwünschten] Einheiten [...] `kürzen´ sich gegenseitig `heraus´ [...] Eine andere Art diesen Vorgang zu beschreiben ist diese. Die [ursprüngliche] Größe [...] ist mit einem Bruch multipliziert worden, dessen Wert 1 ist [...] und Multiplikation mit der Zahl 1 erzeugt keine Veränderung. Dies ist die wahre Bedeutung des Fachausdrucks `*Umrechnungsfaktor*´, einer Größe, die gleich 1 ist und die als Multiplikationsfaktor eingeführt werden kann zu dem einzigen Zweck, Einheiten zu ändern (Romer 1976: 36, kursiv im Original, übersetzt durch SK).

CU2_3_d: In folgendem Beispiel, erdacht von SK, multiplizieren wir die ursprüngliche Größe mit zwei Faktoren von 1, wobei jeder der beiden geeignet ausgedrückt ist als Bruch mit der unerwünschten Einheit in einer solchen Position, dass die unerwünschten Einheiten der ursprünglichen Größe herausgekürzt werden.

$$36\frac{km}{h}=36\frac{km}{h}\cdot1\cdot1=36\frac{km}{h}\cdot\frac{h}{3600s}\cdot1=36\frac{km}{h}\cdot\frac{h}{3600s}\frac{1000m}{km}=36\frac{m}{s}\cdot\frac{1000}{3600}=10\frac{m}{s}$$

ME1: Measurements, Accuracy, Precision / Messungen, Treffgenauigkeit und Wiederholungsgenauigkeit

ME1_1_e: „[*A*]*ccuracy* refers to the *closeness* of the measurements *to the true value* - closeness to some reference or standard value accepted as the truth - whereas *precision* refers merely to their *closeness together*. Thus, *accuracy* expresses a relation to a value *external* to the measurement process; *precision*, to a value *internal* to the process". (Eisenhart 1952: 546, italics in the source; Topping 1963: 14, Topping (1963) uses Eisenhart´s text with only minor changes and without quotation marks.).

ME1_2_e: "An *accurate* method of measuring some quantity is, therefore, a method that is both *precise* and *unbiased*, in the sense that it yields measurements that are *closely clustered* and *centered on the true value*" (Eisenhart 1952: 546, italics in the source).

ME1_3_e: "Accuracy therefore includes precision but the converse is not necessarily true" (Topping 1963: 14).

ME1_4_e: The *precison* of a measurement is related to random errors. The *trueness* of a measurement is related to systematic errors. A high *accuracy* of a measurement results from small random errors as well as small systematic errors (following Mortimer & Müller 2015: 30).

ME1_5_e: The result of the measurement of a quantity is reported as a best estimate and a probable (estimated) range within which the quantity lies (Taylor 1982: 5-6).

ME1_1_d: „*Treffgenauigkeit* [Genauigkeit] bezieht sich auf die *Nähe* der gemessenen Werte *zum wahren Wert* - Nähe zu einem Bezugswert oder Standardwert, der als die Wahrheit akzeptiert ist - während *Wiederholungsgenauigkeit* [Präzision] sich lediglich auf die *Nähe* der gemessenen Werte *zueinander* bezieht. Somit drückt *Treffgenauigkeit* eine Beziehung zu einem Wert *außerhalb* des Prozesses der Messung aus; *Wiederholungsgenauigkeit* zu einem Wert *innerhalb* des Prozesses" (Eisenhart 1952: 546, Kursivschrift im Original; Topping 1963: 14, Topping (1963) verwendet Eisenharts Text mit nur geringfügigen Änderungen und ohne Anführungszeichen. Die Übersetzung von SK folgt daher auch Topping 1975: 7.)

ME1_2_d: „Eine *genaue* Methode, eine bestimmte Größe zu messen, ist daher eine Methode die sowohl *präzise* als auch *unverzerrt* ist in dem Sinne, dass sie Messwerte liefert, die *dicht gehäuft* und *um den wahren Wert zentriert* sind" (Eisenhart 1952: 546, Kursivschrift im Original, übersetzt von SK).

ME1_3_d: „Die Treffgenauigkeit [Genauigkeit] schließt deshalb Wiederholungsgenauigkeit [Präzision] ein, aber die Umkehrung ist nicht notwendig auch wahr" (Topping 1975: 7).

ME1_4_d: Die *Präzision* (Wiederholungsgenauigkeit) einer Messung bezieht sich auf zufällige Fehler. Die *Richtigkeit* einer Messung bezieht sich auf systematische Fehler. Eine hohe *Genauigkeit* ergibt sich, wenn sowohl zufällige Fehler als auch systematisch Fehler klein sind (in Anlehnung an Mortimer & Müller 2015: 30).

ME1_5_d: Als Ergebnis der Messung einer Größe gibt man einen Bestwert (besten Schätzwert) an und einen wahrscheinlichen (geschätzten) Wertebereich, in welchem der wahre Wert der Größe liegt (Taylor 1988: 5).

ME2: Measurements and Uncertainties / Messungen und Unsicherheiten

ME2_1_e: „There are two types of experimental situation that commonly occur. In the first, repeated measurements of a quantity are made where there is no reason to believe that the quantity is varying" (Kirkup 1994: 13).

ME2_2_e: "In the other situation, we are trying to establish the relationship between two quantities - call them A and B. This we would do by changing A and observing the effect this has had on B" (Kirkup 1994: 13). A "systematic variation of an independent variable" (Beller 2004: 125, übersetzt von SK) is one of the distinguishing features of an experiment. Experiment are used to test physical laws, "mathematical relationships between such quantities (equations between quantities)" (Bormann 1978: 6, emphasis removed, translated by SK). Giancoli uses "equations between quantities" and "Newton's second law, $F=m\cdot a$", as an example (Giancoli 2023:25, italics in the source).

ME2_3_e: "Precision is a measure of repeatability" (Adams 2003: 4, boldface and italics removed). "Accuracy is a measure of how close a measurement is to the true value" (Adams 2003: 3, boldface and italics removed).

ME2_4_e: "The precision of a result or method are in no way related to its accuracy" (Drucker 1925: 2, überstzt von SK). "That the result is accurate, even if only approximately, means that it contains no systematic errors" (Drucker 1925: 2, überstzt von SK).

ME2_5_e: There is an "experimental uncertainty (also known as experimental error) in each quantity" (Kirkup 1994: 6) measured.

ME2_1_d: "Es gibt zwei Arten einer experimentellen Situation, die gewöhnlich vorkommen. In der ersten werden wiederholte Messungen einer Größe durchgeführt, wobei kein Grund zur Annahme besteht, dass diese Größe veränderlich ist" (Kirkup 1994: 13, überstzt von SK).

ME2_2_d: „In der anderen Situation versuchen wir, die Beziehung zwischen zwei Größen nachzuweisen - nenen wir sie A und B. Wir tun dies gewöhnlich. indem wir A ändern und beobachten, welche Auswirkung dies auf B gehabt hat" (Kirkup 1994: 13, übersetzt von SK). Eine „systematische Variation einer unabhängigen Variablen" (Beller 2004:125) ist eines der Merkmale eines Experimentes. Damit prüft man unter anderem Physikalische Gesetze, „mathematische Zusammenhänge zwischen solchen Größen (Größengleichungen)" (Bormann 1978: 6, Hervorhebung entfernt). Giancoli benutzt [...] "das zweite Newtonsche Axiom, $F=m\cdot a$", als Beispiel (Giancoli 2023:25, kursiv im Original, übersetzt von SK).

ME2_3_d: „Präzision ist ein Maß für Wiederholbarkeit" (Adams 2003: 4, Fettdruck und Kursivschrift entfernt, übersetzt von SK). „Treffgenauigkeit ist ein Maß dafür, wie nahe ein Messwert am wahren Wert liegt" (Adams 2003: 3, Fettdruck und Kusivschrift entfernt, Übersetzung von SK).

ME2_4_d: „Die Schärfe oder Präzision eines Resultates oder einer Methode stehen in keiner Beziehung zu der Richtigkeit" (Drucker 1925: 2). „[D]aß das Resultat, wenn auch nur angenähert, richtig ist, [...] heißt, daß es keine `systematischen´ Fehler enthält" (Drucker 1925: 2).

ME2_5_e: Es gibt eine „experimentelle Unsicherheit (auch als experimenteller Fehler bekannt) in jeder Größe" (Kirkup 1994: 6, übersetzt von SK), die gemessen wird.

ME3: Measurements Seen Statistically 1 / Messungen aus statistischer Sicht 1

ME3_1_e: The repeated measurement of a quantity X can be interpreted as the drawing of random values x_i from a sample space characterizing the experimental set-up and procedure. The real value of the quantity X is called μ and is unknown in principle. What is reported as the result of the measurement is $\bar{x}$ ("ecks bar´" (Phillips 2000: 97)), the arithmetic mean of n measurements of the quantity X.

ME3_2_e: The larger the number n of single measurements, the closer $\bar{x}$ gets to μ, the real value of X. The value of $\mu_{\bar{x}}$ ("pronounced `mew sub ecks bar´" (Phillips 2000: 97)), the expected value value of the arithmetic means of the samples of n measurements each, equals μ, the unknown real value of the quantity X.

ME3_3_e: According to the central limit theorem, the larger the sample size n, the closer the distribution of the sample averages $\bar{x}$ gets to to a Gaussian normal distribution, centered around μ and with a standard deviation ["sigma ecks bar"] of $\sigma_{\bar{x}} = \frac{\sigma}{\sqrt{n}}$, if σ is the standard deviation of X. This is independent of the original distribution of the single measurements x_i of X. If n equals 30 or more, the distribution of the sample means $\bar{x}$ is already close to a Gaussian normal distribution, whatever the original distribution of the x_i. I have discovered two books so far (Nachtigall & Wirtz 1998: 92; Hughes & Hase 2010. 32) which feature simple and clear illustrations of the central limit theorem.

ME3_1_d: „Die wiederholte Messung einer Größe X kann interpretiert werden als das Ziehen von Zufallsvariablen aus einer Grundgesamtheit, die den Versuchsaufbau und die Messmethode charakterisiert. Der wahre Wert der Größe X wird μ genannt und ist prinzipiell unbekannt. Berichtet wird als Messergebnis $\bar{x}$ („iks quer"), der arithmetische Mittelwert von jeweils n Messungen der Größe X. Dessen Erwartungswert ist μ.

ME3_2_d: „Je größer die Anzahl n der Einzelmessungen ist, desto näher liegt $\bar{x}$ bei μ, dem wahren Wert der Größe X. Der Wert von $\mu_{\bar{x}}$ („mü iks quer"), der Erwartungswert des Mittelwertes des arithmetischen Mittels von jeweils n Messungen der Größe X, ist gleich μ, den unbekannten realen Wert der Größe X.

ME3_3_d: „Nach dem Zentralen Grenzwertsatz nähert sich die Verteilung der Stichprobenmittelwerte $\bar{x}$ einer Gaußschen Normalverteilung zentriert um μ und mit einer Standardabweichung („sigma iks quer") von $\sigma_{\bar{x}} = \frac{\sigma}{\sqrt{n}}$, falls σ die Standardabweichung der Verteilung der Größe X ist. Falls n gleich 30 ist oder größer, liegt die Verteilung der Stichprobenmittelwerte $\bar{x}$ schon nahe einer Gaußschen Normalverteilung, unabhängig von der ursprünglichen Verteilung der x_i in der Grundgesamtheit. Ich habe bislang zwei Bücher entdeckt (Nachtigall & Wirtz 1998: 92; Hughes & Hase 2010. 32), die eine einfache und klare Illustration des Zentralen Grenzwertsatzes zeigen.

[Note: I wrote this page from memory, drawing upon what I had read in four books (Taylor 1982; Taylor 1988; Nachtigall & Wirtz 1998; Hughes & Hase 2010), just to have a draft for the first edition of this booklet. The text probably needs to be be revised for the second edition of this booklet.]

ME4 Measurements Seen Statistically 2 / Messungen aus statistischer Sicht 2

ME4_1_e: The results of a measurement are reported as "Best Estimate ± Uncertainty" (Taylor 1982: 14). With very few repeated measurments, say 4, the uncertainty can be estimated as half the difference of the highest and the lowest value measured, while the average of the four measurements is taken as the best estimate (Taylor 1982: 11-12).

ME4_2_e: The single measurements of a quantity *X* are assumed to be normally dustributed. After a large number of single measurements, their arithmetic mean $\bar{x}$ can be assumed to be very close to the real value of the quantity *X*. If σ is the standard deviation of the single measurements, then it is customary to state the result as x = $\bar{x} \pm \sigma$.

ME4_3_e: The probability that the next single measurement of the quantity *X* (using the same apparatus and the same procedure) lies between $\bar{x} - \sigma$ and x= $\bar{x} + \sigma$ is about 68% (normal distribution). The result of such a measurement is, e.g., x = (1203.25 ± 0.47)m. The value 0.47m is one standard deviation of the distribution of single measurements.

ME4_4_e: The probability that the arithmetic mean $\bar{x}$ of the next 10 measurements of the quantity *X* (using the same apparatus and the same procedure) lies between 1203.25m - $\sigma / \sqrt{10}$ and 1203.25m + $\sigma / \sqrt{10}$ is about 68%. Taking the values from the example above, 0.47m/$\sqrt{10}$ =0.15m. The result of such a measurement is stated as, e.g., x = (1203.25 ± 0.15)m. It needs to be added that $\sigma_{\bar{x}} \approx 0.15m$ is one standard deviation of the mean of ten single measurements (or one standard error). Taking the mean of 10 single measurements reduces the uncertainty of the reported result by a factor of $\sqrt{10} \approx 3$ (Taylor 1982).

ME4_1_d: Das Ergebnis einer Messung wird berichtet als "Bestwert ± Unsicherheit" (Taylor 1988: 13). Bei sehr wenigen Messwiederholungen, sagen wir 4, schätzt man die Messunsicherheit als halbe Differenz des höchsten und des niedrigsten gemessenen Einzelwertes, während der Mittelwert der vier Einzelmessungen als Bestwert (bestmöglicher Schätzwert) genommen wird (Taylor 1988: 13-14).

ME4_2_d: Nehmen wir an, dass die einzelnen Messwerte einer Größe *X* normalverteilt sind. Nach eine großen Anzahl von Einzelmessungen liegt deren Mittelwert $\bar{x}$ vermutlich sehr nahe am wahren Wert der Größe X. Wenn σ die Standardabweichung der Einzelmessungen ist, ist es üblich, das Ergebnis als x = $\bar{x} \pm \sigma$ anzugeben.

ME4_3_d: Die Wahrscheinlichkeit, dass die nächste Einzelmessung von *X* (mit derselben Apparatur und Prozedur) zwischen $\bar{x} - \sigma$ und x= $\bar{x} + \sigma$ liegt, ist ca. 68% (Normalverteilung). Das Ergebnis einer solchen Messung ist z. B., x = (1203,25 ± 0,47)m. Dabei ist der Wert 0,47m eine Standardabweichung der Verteilung der Einzelmessungen.

ME4_4_d: Die Wahrscheinlichkeit beträgt ungefähr 68%, dass der Mittelwert $\bar{x}$ der nächsten zehn Einzelmessungen von *X* (mit demselben Apparat und derselben Prozedur) zwischen 1203,25m - $\sigma / \sqrt{10}$ und 1203,25m + $\sigma / \sqrt{10}$ liegt. Mit den Werten von oben gilt 0,47m/$\sqrt{10}$ =0,15m. Das Ergebnis einer solchen Messung wird angegeben als, z.B., x = (1203,25 ± 0,15)m. Es ist hinzuzufügen, dass $\sigma_{\bar{x}} \approx 0{,}15m$ eine Standardabweichung des Mittelwertes von 10 EInzelmessungen (oder ein Standardfehler) ist. Die Bildung des Mittelwertes von 10 Einzelmessungen vermindert die Unsicherheit des berichteten Ergebnisses um den Faktor $\sqrt{10} \approx 3$ (Taylor 1988).

SF1: Significant figures 1 / Signifikante Stellen 1

SF1_1_e: “In a measurement, a significant figure is a digit that is either reliably known or closely estimated” (Malone e.a. 2013: 15).

SF1_2_e: “The significant figures of a numerical value are all those figures “which are reliably known plus one figure which is estimated” (Mortimer & Müller 2015: 29, übersetzt von SK).

SF1_3_e: An example (Malone e.a. 2013: 15): If three experts estimate the size of the audience of a rock concert to be 11000, 12000 and 13000, the size of the audience was 12000 ± 1000. Thus, the 1 and the 2 are significant figures, the three zeros are not.

SF1_4_e: “The result of calculations can never be more accurate than the least accurate number used to obtain the result” (Mortimer & Müller 2015: 29, translated by SK).

SF1_5_e: Using a folding rule, I obtain 160cm and 80cm. for the length and the width of my desk. $160\text{cm}\times80\text{cm}=12800\text{cm}^2=1.2800\text{m}^2$ is the area of my desk. However, 80cm only has two significant figures, so that the area has to be stated to within two significant figures by rounding: $A=1.3\text{m}^2$. The change of unit makes the number of significant figures easy to recognize without any additional comment.

SF1_6_e: “The precision of a measurement is expressed by the number of figures used to record its value” (Mortimer & Müller 2015: 29, translated by SK).

SF1_7_e: Take the distance of 3.12578km as an example. “That last digit implies that the [...] [distance] is known to within [...] [0.5 centimeters] (Weinstein 2012: 345).

SF1_1_d: “Bei einem Messwert ist eine signifikante Stelle eine Ziffer, die entweder zuverlässig bekannt oder scharf geschätzt ist“ (Malone e.a. 2013: 15, übersetzt von SK).

SF1_2_d: Signifikante Stellen eines Zahlenwertes sind alle Stellen, „die mit Sicherheit bekannt sind, plus eine Stelle, die geschätzt ist“ (Mortimer & Müller 2015: 29)

SF1_3_d: Ein Beispiel (Malone e.a. 2013: 15): Wenn drei Experten die Größe des Publikums eines Rockkonzertes auf 11000, 12000 und 13000 schätzen, war die Größe 12000 ± 1000. Daher sind die 1 und die 2 signifikante Stellen, aber nicht die Nullen.

SF1_4_d: „Die Genauigkeit eines Messergebnisses wird durch durch die Anzahl der Ziffern zum Andruck gebracht, mit der das Ergebnis aufgezeichnet wird“ (Mortimer & Müller 2015: 29).

SF1_5_d: Mit einem Zollstock erhalte ich 160cm und 80cm für die Länge und die Breite meines Schreibtisches. $160\text{cm}\cdot80\text{cm}=12800\text{cm}^2=1{,}2800\text{m}^2$ ist die Fläche meines Schreibtisches. Aber 80cm hat nur zwei signifikate Stellen, so dass die Fläche durch Runden auf zwei Stellen genau angegeben werden muss: $A=1{,}3\text{m}^2$. Der Wechsel der Maßeinheit erlaubt es, die Zahl der signifikanten Stellen ohne Kommentar zu erkennen.

SF1_6_d: “Bei Berechnungen kann das Ergebnis nie genauer sein als die ungenaueste Zahl, die in das Ergebnis eingeht” (Mortimer & Müller 2015: 29).

SF1_7_d: Nehmen wir die Entfernung von 3,12578km als Beispiel. “Die letzte Dezimalstelle läst darauf schließen, dass die [...] [Entfernung] auf [...] [0.5 Zentimeter] genau bekannt ist (Weinstein 2012: 345, übersetzt von SK).

AV1: Approximate Values 1 / Näherungswerte 1

AV1_1_e: "One of the arts of good applied science is to achieve accuracy *adequate for a particular purpose*. The pursuit of excessive accuracy is both time-consuming and costly" (Acton & Squire 1985: 9, italics in the source).

AV1_2_e: "It is only in the final stages of development, or in the routine design of devices that are well understood that precise numerical values may be needed to optimise the design or to fill in details which are not made clear by approximate treatment" (Acton & Squire 1985: 10).

AV1_3_e: "In this book an approximation will be called useful, if the error is expected to lie within the limits $\pm 30\%$, which means a typical error of around 10% (Acton & Squire 1985: 10).

AV1_4_e: "[A]pproximate solutions can strip away the overlying detail to show the essential relationships between the physical variables" (Acton & Squire 1985: 1).

AV1_5_e: In this case we conclude: "An approximate solution [...] is *better* than an exact solution" (Mahajan 2022: 887, italics in the source).

AV1_6_e: "Indeed, the cleverness of modern science consists largely in the ease with which complicated systems can be described in simple terms or by simple models. Its flair consists in knowing when simple models are a good guide to understanding and when they are insufficient" (Maddox 1972: 142)

AV1_1_d: "Einer der Kunstgriffe guter angewandter Wissenschaft ist es, eine Genauigkeit zu erzielen, die *für einen bestimmten Zweck ausreichend* ist. Das Anstreben einer übermäßigen Genauigkeit ist sowohl zeitraubend als auch kostspielig" (Acton & Squire 1985: 9, italics in the source, übersetzt von SK).

AV1_2_d: „Nur in den Endstadien der Entwicklung oder im routinemäßign Entwurf von Geräten, die man gut versteht, könnte man präzise Zahlenwerte benötigen, um den Entwurf zu optimieren oder Details nachzuliefern, die von einer angenäherten Behandlung nicht klar dargestellt werden" (Acton & Squire 1985: 10, übersetzt von SK).

AV1_3_d: „In diesem Buch wird eine Näherungslösung nützlich genannt, wenn ein Fehler innerhalb der Grenzen $\pm 30\%$ erwartet wird, was einen typischen Fehler von 10% bedeutet" (Acton & Squire 1985: 10, übersetzt von SK).

AV1_4_d: „Näherungslösungen können die überlagernden Details abstreifen, um die wesentlichen Beziehungen zwischen den physikalischen Variablen zu zeigen" (Acton & Squire 1985: 1, übersetzt von SK).

AV1_5_d: In diesem Falle schließen wir: „Eine Näherungslösung [...] ist *besser* als eine exakte Lösung" (Mahajan 2022: 887, Kursivschrift im Original, übersetzt von SK).

AV1_6_d: „In der Tat besteht die Raffiniertheit der modernen Wissenschaft weitgehend in der Leichtigkeit, mit der komplexe Systeme mit einfachen Begriffen oder mit einfachen Modellen beschrieben werden können. Das Gespür der Wisenschaft besteht darin zu wissen, wann einfache Modelle ein gute Richtschnur zum Verständnis sind und wann sie unzureichend sind (Maddox 1972: 142, übersetzt von SK).

AV2: Approximate Values 2 / Näherungswerte 2

AV2_1_e: A numerical solution always is an approximate value. "[W]hat is a numerical solution [of an equation]? To say that the answer is 2.54327 correct to five decimal places means that the solution is x=a where a is some number between 2.543265 and 2.543275. Numerical methods [Conte & de Boor 1981; Knorrenschild 2013] do not find a solution; they only put upper and lower bounds on it" (Munroe 1963: 37).

AV2_2_e: error: „in applied mathematics, the difference between a true value and an estimate, or approximation, of that value" (Hoiberg & Levy 2010: 548)

AV2_3_e: "For $a \approx b$ we say a is approximately equal to *b*" (Arens e.a. 2012: 1485, italics in the source, translated by SK).

AV2_4_e: "Find the solution of the equation [...] correct to six decimal places" (Stewart, Redlin & Watson 2002: 365)

AV2_5_e: The symbol "~" (tilde) has the meaning „is very, very roughly equal to" (Mahajan 2018: 143).

AV2_6_e: "The value, correct to six figures, is" (Meyberg & Vachenauer 2003: 113, übersetzt von SK).

AV2_7_e: You may wonder how the Office for National Statistics of the United Kingdom (or other countries) could know so accurately how many people of a certain age group were alive at a specified date, e.g. "65,110,034" of all ages in "mid-2015" (UK 2016: 465). "These unrounded estimates [...] should not be taken to be accurate to the level of detail provided" (UK 2016: 465).

AV2_1_d: Eine numerische Lösung ist immer ein Näherungswert. „[W]as ist eine numerische Lösung [einer Gleichung]? `Die Antwort ist 2,54327 auf fünf Dezimalstellen genau´ zu sagen, bedeutet dies: Die Lösung ist x=a, wobei a irgendeine Zahl zwischen 2,543265 und 2,543275 ist. Numerische Methoden [Conte & de Boor 1981; Knorrenschild 2013] finden keine Lösung, sie geben nur obere und untere Grenzen der Lösung an" (Munroe 1963: 37, übersetzt von SK).

AV2_2_d: Abweichung: "in der angewandten Mathematik die Differenz zwischen einem wahren Wert und einem Schätzwert oder Näherungswert für diesen Wert" (Hoiberg & Levy 2010: 548, translated by SK)

AV2_3_d: „Für $a \approx b$ sagt man a ist ungefähr gleich b"(Arens e.a. 2012: 1485).

AV2_4_d: „Finde die Lösung der Gleichung [...] auf sechs Dezimalstellen genau" (Stewart, Redlin & Watson 2002: 365, übersetzt von SK)

AV2_5_d: Das Symbol „~" (Tilde) hat die Bedeutung „ist sehr, sehr grob gleich" (Mahajan 2018: 143, übersetzt von SK).

AV2_6_d: „Der auf 7 Stellen genaue Wert ist" (Meyberg & Vachenauer 2003: 113).

AV2_7_d: Sie mögen sich fragen, wie das Statistische Landesamt des Vereinigten Königreiches (oder anderer Länder) so genau wissen kann, wie viele Leute eines bestimmten Alters zu einer angegebenen Zeit lebten, z.B. „65110034" aller Altersgruppen „in der Mitte des Jahres 2015" (UK 2016: 465, übersetzt von SK). "Diese ungerundeten Schätzwerte [...] sollten nicht verstanden werden als genau bis auf die Ebene der gelieferten Details" (UK 2016: 465, übersetzt von SK).

AV3: Approximate Values 3 / Näherungswerte 3

AV3_1_e: "Order of magnitude: This term indicates the relative size of a number [...] Specifically, it refers to the power of ten that can be factored out of a number" (Adams 2003: 2, boldface and italics removed).

AV3_2_e: "A rough estimate [...] made by rounding off all numbers to one significant figure and its power of 10 [...] is called an order-of-magnitude estimate [...] `[O]rder of magnitude´ is sometimes used to refer simply to the power of 10" (Giancoli 2023: 33, boldface removed).

AV3_3_e: Order-of-magnitude estimate: "We are dealing with *round* numbers" (Weinstein & Adams 2008: 58, italics in the source).

AV3_4_e: "When working with approximate values [...] the final answer should be expressed with the same accuracy as the *least*-accurate measurement given in the statement of the problem" (Stewart, Redlin & Watson 2002: xxi, italics in the source.)

AV3_5_e: "The phrase `order of magnitude estimate´ is often used to describe a calculation that is expected to be within a factor of 5 or 10 of the true result, but not necessarily much closer. Two numbers which differ by no more than a factor of 5 or so are said to be `of the same oder of magnitude´"(Romer 1976: 9).

AV3_6_e: "The usual convention is that if we write r=2.5 ft, this means that r is probably closer to 2.5 than to 2.4 or 2.6" (Romer 1976: 9, italics in the source). This implies an accuracy of half a unit of the last digit stated.

AV3_1_d: „Größenordnung: Dieser Ausdruck zeigt die relative Größe einer Zahl an [...] Insbesondere bezieht er sich auf die Zehnerpotenz, die von einer Zahl abgespalten werden kann" (Adams 2003: 2, Fettdruck und Kursivschrift entfernt, übersetzt von SK).

AV3_2_d: „Ein grober Schätzwert [...] erhalten duch Runden aller Zahlen auf eine signifikante Stelle und ihre Zehnerpotenz [...] wird Schätzung der Größenordnung genannt [...] `Größenordnung´ bezieht sich manchmal einfach auf die Zehnerpotenz" (Giancoli 2023: 33, Fettdruck entfernt, übersetzt von SK).

AV3_3_d: Schätzung der Größenordnung: „Wir befassen uns mit *runden* Werten" (Weinstein & Adams 2008: 58, kursiv im Original, überstzt von SK).

AV3_4_d: „Wenn man mit Näherungswerten arbeitet, [...] sollte das Endergebnis mit derselben Genauigkeit angegeben werden wie der *am wenigsten* genaue Messwert in der Darstellung der Aufgabe" (Stewart, Redlin & Watson 2002: xxi, kursiv im Original, übersetzt von SK.)

AV3_5_d: „Der Ausdruck `Schätzung der Größenordnung´ wird häufig verwendet, um eine Rechnung zu bescheiben, von der man erwartet, dass sie innerhalb eines Faktors von 5 oder 10 des wahren Wertes liegt, aber nicht notwedigerweise viel näher. Von zwei Zahlen, die sich nicht um mehr als einen Faktor von etwa 5 unterscheiden, sagt man, dass sie `von derselben Größenordnung´ sind" (Romer 1976: 9, übersetzt von SK).

AV4_6_d: „Nach der üblichen Konvention bedeutet es, r=2,5 ft zu scheiben, dass r wahrscheinlich näher an 2,5 liegt als an 2,4 oder 2,6" (Romer 1976: 9, kursiv im Original, übersetzt von SK). Dies deutet auf eine Genauigkeit von einer halben Einheit der letzten angegebenen Stelle hin.

MM1: Mathematical models 1 / Mathematische Modelle 1

MM1_1_e: "In endeavoring this to draw inferences as to the future from the past, we always adopt the following process. We form for ourselves images or symbols of external objects ; and the form which we give them is such that the necessary consequents of the images in thought are always the images of the necessary consequents in nature of the things pictured. In order hat this requirement may be satisfied, there must be a certain conformity between nature and our thought. Experience teaches us [...] that such a conformity in fact exists. [...] The images we speak of are our conceptions of things. With the things themselves they are in conformity in one important respect, namely, in satisfying the above-mentioned requirement"(Hertz 1899: 1).

MM1_2_e: "A mathematical model is a mathematical representation of an object or process. Often a mathematical model is a function that describes a certain phenomenon. [...] Mathematical models are useful because they enable us to isolate critical aspects of the thing we are studying and then to predict how it will behave. [...] Aircraft manufacturers use elaborate mathematical models to predict the aerodynamic properties of a new design *before* the aircraft is actually built" (Stewart, Redlin & Watson 2002: 246, italics in the source, boldface removed).

MM1_1_d: „Das Verfahren aber, dessen wir uns zur Ableitung des Zukünftigen aus dem Vergangenen und damit zur erstrebten Voraussicht stets bedienen, ist dieses: Wir machen uns innere Scheinbilder oder Symbole der äußeren Gegenstände, und zwar machen wir sie von solcher Art, daß die denknotwendigen Folgen der Bilde stets wieder die Bilder seien von den naturnotwendigen Folgen der abgebildeten Gegenstände. Damit diese Forderung überhaupt erfüllbar sei, müssen gewisse Übereinstimmungen vorhanden sein zwischen der Natur und unserem Geiste. Die Erfahrung lehrt uns, daß diese Forderung erfüllbar ist und daß solche Übereinstimmungen in der Tat bestehen. [...] (Hertz 1894: 1). „Die Bilder, von welchen wir reden, sind unsere Vorstellungen von den Dingen; sie haben mit den Dingen die eine wesentliche Übereinstimmung, welche in der Erfüllung der genannten Forderung liegt"(Hertz 1894: 2).

MM1_2_e: „Ein mathematisches Modell ist eine mathematische Repräsentation eines Gegenstandes oder eines Vorgangs. Häufig ist ein mathematisches Modell eine Funktion, die eine bestimmte Erscheinung beschreibt. [...] Hersteller von Flugzeugen benutzen ausgefeilte mathematische Modelle, um die aerodynamischen Eigenschaften eines neuen Flugzeugentwurfes vorherzusagen, *bevor* das Flugzeug tatsächlich gebaut wird" (Stewart, Redlin & Watson 2002: 246, Kursivschrift im Original, Fettdruck entfernt, übersetzt von SK).

MM2: Mathematical models 2 / Mathematische Modelle 2

MM2_1_e: Simple modelling with functions of one variable involves four steps. The quantity to be modelled should be expressed in words "as a function of the other quantities in the problem" (Stewart, Redlin & Watson 2002: 209). The next step is to "[a]ssign a symbol, such as *x*, to one variable and [to] express the other variables in terms of this symbol" (Stewart, Redlin & Watson 2002: 209, italics in the source). The third step is to write the function established in step 1 in terms of *x*. In the fourth step, a mathematical procedure is applied to this function "to answer the questions posed in the problem" (Stewart, Redlin & Watson 2002: 209).

MM2_2_e: These four steps are illustrated by describing the problem what the radius and the height of a can of 250cm³ volume need to be so that the cost of materials for the production of a tin is a minimum. The amount of metal needed is proportional to the surface area of the tin. (1) The surface area is the sum of the areas of the bottom surface, the upper surface and the sides. (2) We assign the letter r to the radius of the can. The bottom surface and the upper surface each have the area $\pi \cdot r^2$. The area of the sides can be written as $2 \cdot \pi \cdot r \cdot h$, with *h* assigned to the height of the tin. This can be expressed in terms of r by solving the formula for the volume of the tin, $250cm^3 = \pi \cdot r^2 \cdot h$, for *h*. (3) We obtain $A(r) = 2 \cdot \pi \cdot r^2 + 2 \cdot \pi \cdot r \cdot (250cm^3 / \pi \cdot r^2) = 2 \cdot \pi \cdot r^2 + (500cm^3 / r)$ for the size of the surface area *A* of the tin. (4) We calculate the first derivative $A'(r)$ of the function $A(r)$, set this equal to 0 and obtain r=3,41cm and h=6,83cm.

MM2_1_d: Das einfache Modellieren mit Funktionen einer Variablen beinhaltet vier Schritte. Die zu modellierende Größe soll ausgedrückt werden „als eine Funktion der anderen Größen in dem Problem“ (Stewart, Redlin & Watson 2002: 209, übersetzt von SK). Der nächste Schritt besteht darin, „ein Symbol, zum Beispiel *x*, einer der Variablen zuzuweisen und die anderen Variablen durch dieses Symbol auszudrücken“ (Stewart, Redlin & Watson 2002: 209, Kursivschrift im Original, übersetzt von SK). Der dritte Schritt besteht darin, die in Schritt 1 aufgestellte Funktion mit Hilfe von *x* zu schreiben. Im vierten Schrittt wird eine mathematische Prozedur auf diese Funktion angewandt, „um die im Problem gestellten Fragen zu beantworten“ (Stewart, Redlin & Watson 2002: 209).

MM2_2_d: Diese vier Schritte werden anhand des Problems dargestellt, welchen Radius und welche Höhe eine Konservendose von 250cm³ Inhalt haben muss, bei der der Materialverbrauch zur Herstellung minimal ist. Der Metallverbrauch ist proportional zur Oberfläche. Diese muss minimal sein. (1) Die Oberfläche ist die Summe aus Grundfläche, Deckfläche und Mantelfläche. (2) Wir weisen dem Radius der Dose den Buchstaben r zu. Grundfläche und Deckfläche sind $\pi \cdot r^2$. Die Mantelfläche ist $2 \cdot \pi \cdot r \cdot h$, wobei *h* die Höhe der Dose ist. Diese drücken wir durch *r* aus, indem wir die Formel zur Berechnung des Dosenvolumens, $250cm^3 = \pi \cdot r^2 \cdot h$, nach *h* auflösen. (3) Wir erhalten $A(r) = 2 \cdot \pi \cdot r^2 + 2 \cdot \pi \cdot r \cdot (250cm^3 / \pi \cdot r^2) = 2 \cdot \pi \cdot r^2 + (500cm^3 / r)$ für die Größe der Oberfläche. (4) Von dieser Funktion $A(r)$ bestimmen wir die erste Ableitung $A'(r)$, setzen diese gleich 0 und erhalten damit r=3,41cm sowie h=6,83cm.

MM3: Mathematical models 3 / Mathematische Modelle 3

MM3_1_e: "Applied mathematics involves establishing an equivalence between the empirical, real-world context (*target domain*) and abstract mathematical structures (*source domain*). To do so, we need to idealize away from the messiness of the real world with models. After `doing the math´ with the help of mathematical theory in the source domain, the derived mathematical truths are interpreted back in the target domain via our models, where we can create predictions, representations, and descriptions" (Barboianu 2022: 91, italics in the source).

MM3_2_e: In our example of the can with the smallest cost of materials mathematics is the source domain, to be more specific, geometry and differential calculus. The target domain is the real world of cans. We disregard offcuts and the seams. Furthermore, we do not consider further requirements our can might have to meet. E.g., we could have demanded a radius small enough to make the can easy to grab. In the source domain of mathematics we use geometry to calculate volume and surface area of cylinders and we use differential calculus to determine the maximum or minimum of functions. With the results obtained in this domain, r=3,41cm und h=6,83cm, we can produce cans in the target domain, which have a volume of 250cm³ and the smallest possible cost of materials.

MM3_1_d: "Angewandte Mathematik beinhaltet, eine Äquivalenz zu bilden zwischen einem empirischen Zusammenhang in der realen Welt (*Zieldomäne*) und abstrakten mathematischen Strukturen (*Quellendomäne*). Um dies zu tun, müssen wir mit Hilfe von Modellen von der Unordentlichkeit der realen Welt abstrahieren. Nachdem wir dann in der Quellendomäne mit Hilfe mathematischer Theorie „die Mathematik getrieben haben", werden die abgeleiteten mathematischen Wahrheiten über unsere Modelle zurück in der Zieldomäne interpretiert, wo wir Vorhersagen, Repräsentationen und Beschreibungen erzeugen können" (Barboianu 2022: 91, Kursivschrift im Original, übersetzt von SK).

MM3_2_d: In unserem Beipiel der materialsparenden Konservendose ist die Quellendomäne die Mathematik, genauer die Geometrie und die Differentialrechnung. Die Zieldomäne ist die reale Welt der Konservendosen. Wir vernachlässigen den Verschnitt und die Materialbedarf für die Falze der Dose. Darüberhinaus betrachten wir keine weiteren Anforderungen an unsere Dose. Man hätte zum Beispiel einen so kleinen Durchmesser fordern können, dass sich die Dose bequem greifen lässt. Wir verwenden in der Quellendomäne, der Mathematik, die Geometrie zur Berechnung von Volumen und Oberfläche von Zylindern und die Methode, durch Bilden der ersten Ableitung Maxima und Minima von Funktionen zu finden. Mit den in dieser Domäne erzielten Ergebnissen, r=3,41cm und h=6,83cm, können wir in der Zieldomäne materialsparende Konservendosen Dosen von 250cm³ Volumen herstellen.

CS1: Cartesian coordinate system 1 / Cartesisches Koordinatensystem 1

CS1_1_e: "Two perpendicular reference lines (called *coordinate axes*) are chosen, one horizontal (called the `x-axis´), the other vertical (called the `*y*-axis´). Their point of intersection, denoted by [...] [*O*(0|0) (German) or *O*(0,0) (English)] is called the *origin*" (Apostol 1967: 48, italics in the source).

CS1_2_e: "Each point in the plane (sometimes called *xy*-plane) is assigned a pair of numbers, called its *coordinates*" (Apostol 1967: 48, italics in the source). The form "a unique ordered pair of numbers (*a*,*b*)" (Stewart, Redlin & Watson 2002: 90, boldface removed, italics in the source) or in German texts (*a*;*b*). The frst coordinate of the point *Q*, *a*, is the distance of the point from the y-axis. A negative sign is placed in front of the number if the point lies to left of the y axis. The second coordinate, *b*, is the distance of the point from the x-axis. A negative sign is placed in front of the number if the point lies below the x-axis. For example, the point *P*(-5,4), or *P*(-5|4) in German texts, is plotted 5 units to the left of the *y*-axis and 4 units above the *x*-axis (Apostol 1967: 48).

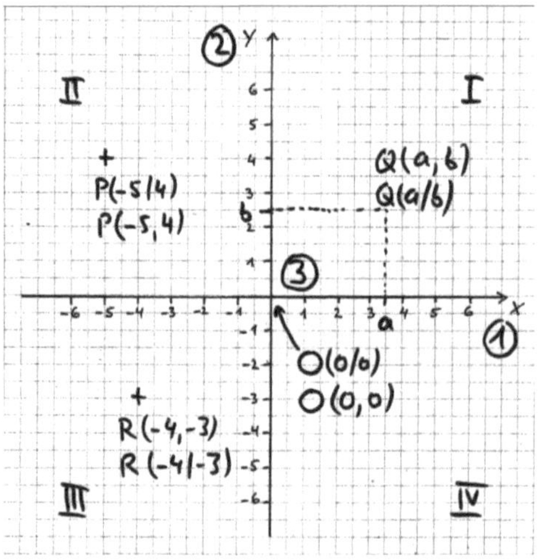

Figure: Points in a Cartesian coordinate system

Abbildung: Punkte in einem Cartesischen Koordinatensystem

CS1_1_d: „Zwei senkrechte Bezugslinien (*Koordinatenachsen* genannt) werden gewählt, eine horizontal (`*x*-Achse´ genannt), die andere vertikal (`*y*-Achse´ genannt). Ihr Schnittpunkt, bezeichnet mit [...] [*O*(0|0) (Deutsch) oder *O*(0,0) (Englisch)] wird *Ursprung* genannt" (Apostol 1967: 48, kursiv im Original, übersetzt von SK).

CS1_2_d: „Jedem Punkt in der Ebene (manchmal *xy*-Ebene genannt) wird ein Zahlenpaar zugeordnet, seine *Koordinaten* (Apostol 1967: 48, kursiv in Original, übersetzt von SK). Diese bilden „ein eindeutiges, geordnetes Zahlenpaar (*a*,*b*)" oder in Deutschen Texten (*a*;*b*). Die erste Koordinate des Punktes *Q*, *a*, ist der Abstand des Punktes von der *y*-Achse. Ein negatives Vorzeichen wird vor die Zahl gesetzt, wenn der Punkt links von der *y*-Achse liegt. Die zweite Koordinate, *b*, ist der Abstand des Punktes von der x-Achse. Ein negatives Vorzeichen wird vor die Zahl gesetzt, wenn der Punkt unterhalb der *x*-Achse liegt. Zum Beispiel zeichnet man den Punkt *P*(-5,4), oder *P*(-5|4) in Deutschen Texten, 5 Einheiten links von der *y*-Achse und 4 Einheiten oberhalb der x-Achse ein (Apostol 1967: 48).

CS2: Cartesian coordinate system 2 / Cartesisches Koordinatensystem 2

CS2_1_e: "The graph of an equation in *x* and *y* is the set of all points (*x*,*y*) in the coordinate plane that satisfy the equation" (Stewart, Redlin & Watson 2002: 93, italics in the source, boldface removed)

CS2_1_e: The parabola sketched in the coordinate system arises as follows. We calculate $f(-2) = (-2)^2 = 4$, $f(-1) = (-1)^2 = 1$, $f(0) = 0^2 = 0$, $f(1) = 1^2 = 1$, and $f(2) = 2^2 = 4$, and plot the points which have the *x*-values as their first coordinate and the f(x)-values as their second coordinate. If we were to do this for about 100 to 200 points, the resulting curve would be indistinguishable from our sketch drawn with the help of a stencil. Similarly, the sketch of the straight line in the figure was produced by calculating the *y*-coordinates for some *x*-coordinates, e.g. g(-2) = -4. The resulting points, e.g. (-2,-4), were plotted and connected with the help of a ruler.

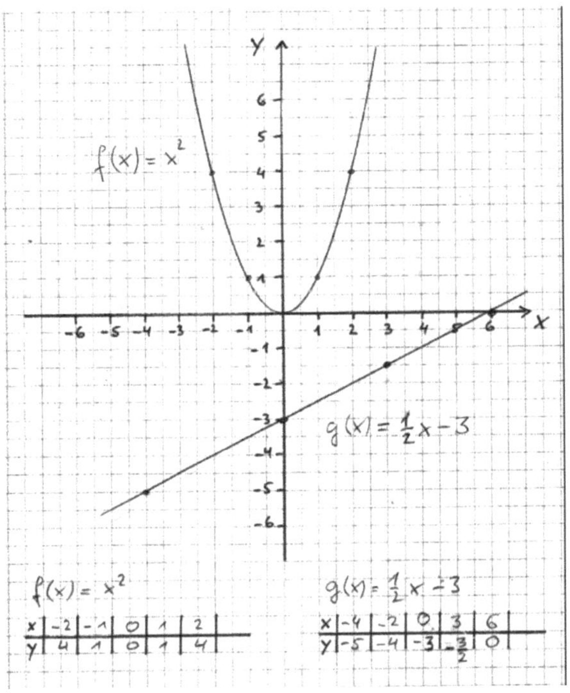

Figure: Graphs in a Cartesian coordinate system

Abbildung: Graphen in einem Cartesischen Koordinatensystem

CS2_1_d: „Der Graph einer Gleichung in x und y ist die Menge aller Punkte (x|y) in der Koordinatenebene, welche die Gleichung erfüllen" (Stewart, Redlin & Watson 2002: 93, italics in the source, boldface removed, übersetzt von SK)

CS2_2_d: Die im Koordinatensystem skizzierte Parabel kommt folgendermaßen zustande. Wir berechnen $f(-2) = (-2)^2 = 4$, $f(-1) = (-1)^2 = 1$, $f(0) = 0^2 = 0$, $f(1) = 1^2 = 1$, und $f(2) = 2^2 = 4$ und tragen die Punkte ins Koordinatensystenm ein, welche die x-Werte als erste Koordinate und die f(x)-Werte als zweite Koordinate haben. Täten wir dies für etwa 100 bis 200 Punkte, wäre die resultierende Kurve ununterscheidbar von unserer Skizze, die mit Hilfe einer Schablone gezeichnet wurde. Ähnlich wurde die Skizze der Gerade in der Abbildung erzeugt, indem die *y*-Koordinaten für einige *x*-Koordinaten berechnet wurden, z.B. g(-2) = 4. Die resultierenden Punkte, z. B, (-2|-4), wurden eingezeichnet und mit Hilfe eines Lineals verbunden.

SL1: Straight Lines 1 / Geraden 1

SL1_1_e: "*y*= *mx*+*b* [...] is called the slope-intercept form of the equation of a line" (Stewart, Redlin & Watson 2002: 117, boldface removed, intalics in the source)

SL1_2_e: As the line passes through the point (0,*b*) on the *y*-axis, *b* is called the intercept, or *y*-intercept in British books (Smedley & Wiseman 2001: 143). The factor *m* is called the slope, or gradient (Smedley & Wiseman 2001: 143) in British books.

SL1_3_e: The slope *m* of the straight line can be calculated as the ratio of rise over run, the step taken in *y*-direction, y_2-y_1, divided by the step taken in *x*-direction, x_2-x_1:

$$m = \frac{rise}{run} = \frac{y_2 - y_1}{x_2 - x_1} = \frac{6-4}{6-2} = \frac{2}{4} = \frac{1}{2} \qquad m = \frac{rise}{run} = \frac{y_2 - y_1}{x_2 - x_1} = \frac{-3-(-2)}{6-3} = \frac{-1}{3}$$

SL1_4_e: In the first case, "for every [...] [2] units we move to the right, the line rises [...] [1] unit" (Stewart, Redlin & Watson 2002: 115). In the second case, "for every [...] [3] units we move to the right, the line rises [...] [-1] unit" (Stewart, Redlin & Watson 2002: 115). So the line drops by one unit and the slope is negative.

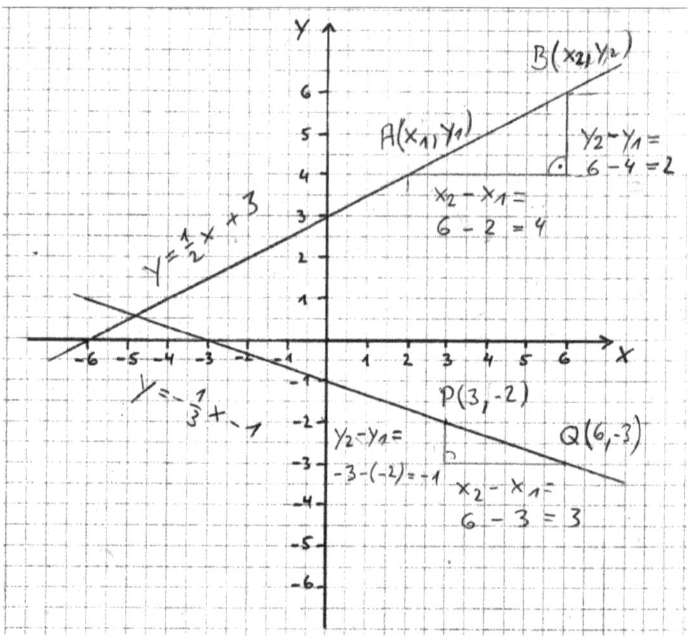

Figure: Graphs of straight lines Abbildung: Graphen von Geraden

SL1_1_d: *y*= *mx*+*b* [...] wird Normalform der Geradengleichung genannt (Kreul & Ziebarth 2026. 491).

SL1_2_d: Weil die Gerade durch den Punkt (0 | b) auf der y-Achse verläuft, heißt der Wert b y-Achsabschnitt. Der Faktor m heißt Steigung.

SL2_3_d: Die Steigung *m* der Geraden ergibt sich als Quotient der Schrittweite in y-Richtung, , y_2-y_1, und der Schrittweite in *x*-Richtung, x_2-x_1. [Siehe oben]

SL2_4_d: Im ersten Fall steigt die Gerade für je zwei Einheiten, die man sich nach rechts bewegt, um eine Einheit an. Im zweiten Fall fällt die Gerade für je drei Einheiten, die man sich nach rechts bewegt, um eine Einheit ab. Die Steigung ist negativ.

SL2: Straight Lines 2 / Geraden 2

SL2_1_e: Two straight lines are sketched in the coordinate system. We evalutae both functions, f(x) and g(x), for some arbitrarily chosen x-values, and plot the resulting points (x,y). For example, $g(-6)=\frac{3}{4}\cdot(-6)+3=-\frac{3}{2}$, g(0) = 3, $g(-2)=\frac{3}{4}\cdot(-2)+3=\frac{3}{2}$, $f(-4)=-\frac{11}{4}\cdot(-4)-4=7$, $f(-2)=-\frac{11}{4}\cdot(-2)-4=\frac{3}{2}$, etc.

SL2_2_e: We see that g(-2) = 3/2 and that f(-2) = 3/2. Both functions have the same y-value if the x-value is -2. The point (-2,3/2) is the point of intersection of their graphs.

SL2_3_e: We could have found this point of intersection by equating the right-hand sides of the equations for f(x) and g(x) and solving for x: $\frac{3}{4}\cdot x+3=-\frac{11}{4}\cdot x-4$, $\frac{3}{4}\cdot x+\frac{11}{4}\cdot x=-4-3$, $\frac{14}{4}\cdot x=-7$, $14\cdot x=-28$, $x=-2$, $f(-2)=\frac{3}{2}$

SL2_4_e: Does the point (-1,-1) lie on the graph of f(x)? It almost looks like that. But f(-1) = -1,25 ≠ -1. Thepoint is not on the graph. The sketch is not accurate enough.

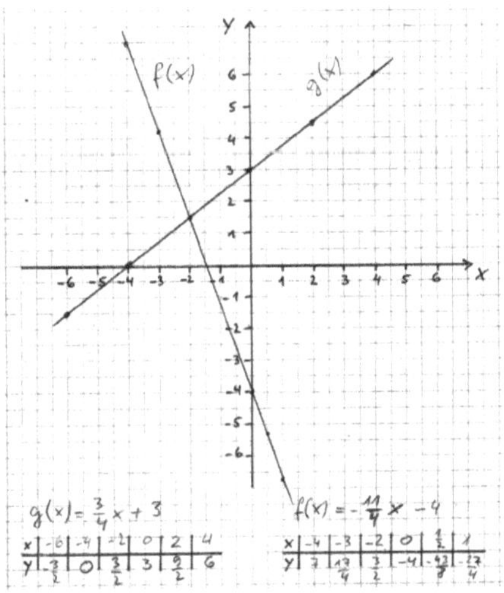

Figure: Graphs in a Cartesian coordinate system

Abbildung: Graphen in einem Cartesischen Koordinatensystem

SL2_1_d: Zwei Geraden werden in einem Koordinatensystem skizzert. Wie berechnen die Funktionswerte beider Funktionen, f(x) und g(x), für einige beliebig ausgewählte x-Werte, und tragen die resultierenden Punkte (x|y) ein. Zum Beispiel, [...] [Siehe oben]

SL2_2_e: Wir sehen, dass g(-2) = 3/2 und dass f(-2) = 3/2. Beide Funktionen haben denselben y-Wert, wenn der x-Wert -2 ist. Der Punkt (-2|3/2) der Schnittpunkt ihrer Graphen.

SL2_3_d: Wir könnten diesen Punkt auch gefunden haben durch Gleichsetzen der rechten Seiten der Gleichungen für f(x) und g(x) und Auflösen nach x. [Siehe oben.]

SL2_4_d: Liegt der Punkt (-1 | -1) auf dem Graphen von f(x)? Es sieht fast so aus. Aber f(-1) = -1,25 ≠ -1. Der Punkt liegt nicht auf dem Graphen. Die Skizze ist nicht genau genug.

LM1: Limits 1 / Grenzwerte 1

LM1_1_e: "$\Delta x \to 0$" "Δx tends to zero" (Bronshtein & Semendyayev 1973: 360)

LM1_2_e: "As x gets very negative, the value of e^x approaches 0, but from above" (Gould & Hurst 2009: 24).

LM1_3_e: "As *x* gets larger, e^x gets larger *incredibly quickly*" (Gould & Hurst 2009: 24, italics in the source).

LM1_4_e: "$n! \sim n^n \cdot e^{-n} \cdot \sqrt{2 \cdot \pi \cdot n}$ [...] Stirling's formula" (Boas 1983: 472) as an approximation for n! [n factorial: n·(n-1)·(n-2)·...2·1]. "The sign ~ (read \`is asymptotic to´) means that the ratio of the two sides $\frac{n!}{n^n \cdot e^{-n} \cdot \sqrt{2 \cdot \pi \cdot n}}$ tends to 1 as $n \to \infty$" (Boas 1983: 472-473).

LM1_5_e: "The expression $\lim\limits_{\Delta t \to 0} \frac{\Delta b}{\Delta t}$ [is] pronounced \`the limit as delta *t* approaches 0 of delta *b* over delta *t* [...]´" (Adler 1998: 124, italics in the source).

LM1_6_e: Those who practice \`deeper mathematics´ (Beck & Geoghegan 2010: iii) prefer "a definition of \`converges´ that does not use words like \`nearer and nearer to´ or \`approaches´"(Beck & Geoghegan 2010: 95).

LM1_1_d: "$\Delta x \to 0$" "Δx geht gegen 0" (Bronshtein & Semendyayev 1973: 360)

LM1_2_d: "Wenn x sehr negativ wird, nähert sich der Wert von e^x der 0, aber von oben" (Gould & Hurst 2009: 24, übersetzt von SK).

LM1_3_d: "Wenn *x* größer wird, wird e^x *unglaublich schnell* größer" (Gould & Hurst 2009: 24, italics in the source).

LM1_4_d: "$n! \sim n^n \cdot e^{-n} \cdot \sqrt{2 \cdot \pi \cdot n}$ [...] Stirlingsche Formel" (Boas 1983: 472) als Näherung für n! [n Fakultät: n·(n-1)·(n-2)·...2·1]. "Das Zeichen [Tilde] ~ (lies \`strebt asymptotisch gegen ´) bedeutet, dass das Verhältnis der beiden Seiten $\frac{n!}{n^n \cdot e^{-n} \cdot \sqrt{2 \cdot \pi \cdot n}}$ gegen 1 geht wenn $n \to \infty$" (Boas 1983: 472-473).

LM1_5_d: "Der Ausdruck $\lim\limits_{\Delta t \to 0} \frac{\Delta b}{\Delta t}$ [wird] ausgesprochen \`der Grenzwert wenn delta *t* sich 0 annähert von delta *b* durch delta *t* [...]´" (Adler 1998: 124, kursiv im Original, übersetzt von SK).

LM1_6_d: Leute, die \`tiefgründigere Mathematik´ (Beck & Geoghegan 2010: iii, übersetzt von SK) betreiben, bevorzugen „eine Definition von \`konvergiert´, die keine Wörter benutzt wie \`näher und näher´ oder \`nähert sich´"(Beck & Geoghegan 2010: 95, übersetzt von SK).

DC1: Derivatives 1 / Ableitungen 1

DC1_1_e: "Differentiation is the process of determining how quickly or slowly a function varies, as the quantity on which it depends, its *argument*, is changed. More specifically it is the procedure for obtaining an expression (numerical or algebraic) for the rate of change of the function with respect to its argument" (Riley, Hobson & Bence 2003: 41, italics in the source).

DC1_2_e: "The derivative $f'(x)$ ["read as `f prime of x´"] is defined by the equation $f'(x) = \lim\limits_{h \to 0} \frac{f(x+h) - f(x)}{h}$, provided the limit exists. The number $f'(x)$ is also called the rate of change of f at x" (Apostol 1967: 160, italics removed).

DC1_3_e: "The quotient itself is referred to as the average rate of change of f in the interval joining x to x+h" (Apostol 1967: 160, italics removed).

DC1_4_e: "Now we let h approach zero and see what happens to this quotient. If the quotient approaches some definite value as a limit (which implies that the limit is the same whether h approaches zero through positive values or through negative values), then this limit is called the derivative of f at x and is denoted by the symbol f'(x) (read as `f prime of x´)(Apostol 1967: 160, italics removed).

DC1_5_e: "If y=f(x) is represented by a graph in Cartesian coordinates [...], then f´(x)=tan(α) [f of x equals tangent of alpha"], where α is the angle between the x axis and the tangent line to the curve in the given point" (Bronshtein & Sem. 1973: 360).

DC1_1_d: „Differentiation ist das Verfahren, mit dem man bestimmt, wie schnell oder langsam eine Funktion ihren Wert ändert, wenn die Größe, von der sie abhängt, ihr *Argument*, verändert wird. Genauer gesagt, ist es die Prozedur dafür, einen Ausdruck (numerisch oder algebraisch) zu erhalten für die Änderungsrate der Funktion bezogen auf ihr Argument" (Riley, Hobson & Bence 2003: 41, Kursivschrift im Original, übersetzt von SK).

DC1_2_d: „Die Ableitung $f'(x)$ ["lies als `f Strich von x´"] ist definiert durch die Gleichung $f'(x) = \lim\limits_{h \to 0} \frac{f(x+h) - f(x)}{h}$, falls der Grenzwert existiert. Die Zahl $f'(x)$ wird auch Veränderungsrate von f an der Stelle x genannt" (Apostol 1967: 160, Kursivschrift entfernt, übersetzt von SK).

DC1_3_d: „Der Quotient selber wird bezeichnet als die mittlere Veränderungsrate von f in dem Intervall, das x mit x+h verbindet" (Apostol 1967: 160, Kursivschrift entfernt, übersetzt von SK).

DC1_4_d: „Now we let h approch zero and see what happens to this quotient. If the quotient approaches some definite value as a limit (which implies that the limit is the same whether h approaches zero through positive values or through negative values), then this limit is called the derivative of f at x and is denoted by the symbol f'(x) (read as `f prime of x´)(Apostol 1967: 160, italics removed, übersetzt von SK).

DC1_5_d: „Falls y=f(x) dargestellt wird durch einen Graphen in cartesischen Koordinaten [...], dann gilt f´(x)=tan(α) [„f von x gleich Tangens von alpha"], wobei α der Winkel zwischen x-Achse und der Tangente an die Kurve im gegebenen Punkt ist" (Bronshtein & Semendyayev 1973: 360, übersetzt von SK).

DC2: Derivatives 2 / Ableitungen 2

DC2_1_e: "[P]rime notation defines a new function $f'(t)$ [`*f* prime of *t*'] that outputs the rate of change of $f(t)$ at any input time t" (Adler 1998: 124, italics in the source).

DC2_2_e: "We can also use the point-slope form of a line [...] to write the *equation* of the tangent line. Suppose the function $f(x)$ has [a] derivative equal to $f'(a)$ at the base point *a*. The tangent line passes through the point (*a*, *f*(*a*)) with slope $f'(a)$. The tangent line, which we can think of the tangent line approximation, $\hat{f}$ ["*f* hat"], has [the] formula $\hat{f}(x) = f'(a)\cdot(x-a)+f(a)$" (Adler 1998: 124, italics in the source, boldface removed), ["*f* hat of *x* equals *f* prime of *a* times [pause] *x* minus *a* [pause] plus *f* of *a*"].

DC2_3_e: The following terms denote analogous notions: average rate of change, difference quotient, average velocity

DC2_4_e: The following terms denote analogous notions: local rate of change, differential coefficient, instantaneous velocity

DC2_5_e: Other expressions for f'(x): $\frac{df}{dx}$ „the derivative of f with respect to x" (Merziger & Wirth 2010: 13, translated by SK) "gradient function" (Smedley & Wiseman 2001:158)

DC2_6_e: If y=*f*(x), then *dy*/*dx* can be thought of either as the slope of the curve *y*=*f*(*x*) or as the rate of change of *y* with respect to *x* (Boas 1983: 145, italics in the source).

DC2_1_d: „Die Strich-Notation definiert eine neue Funktion *f'*(*t*) [*f* Strich von *t*], die die Veränderungsrate von *f*(*t*) zu jeder Eingabezeit *t* ausgibt" (Adler 1998: 124, kursiv im Original, übersetzt von SK).

DC2_2_d: „Wir können auch die Punkt-Richtungs-Form einer Gerade benutzen [...], um die *Gleichung* der Tangente zu schreiben. Nehmen wir an die Funktion *f*(*x*) habe eine Ableitung gleich *f'*(*a*) am Aufpunkt *a*. Die Tangente verläuft durch den Punkt (*a*|*f*(*a*)) mit Steigung *f*'(*a*). Die Tangente, welche wir uns als die Tangentenapproximation [von f(x)] vorstellen können, $\hat{f}$ [„*f* Dach"], hat die Formel ["*f* Dach von *x* gleich *f* Strich von *a* mal [Pause] *x* minus *a* [Pause] plus *f* von *a*"]: $\hat{f}(x) = f'(a)\cdot(x-a)+f(a)$" (Adler 1998: 124, kursiv im Original, Fettdruck entfernt, übersetzt von SK].

DC2_3_d: Die folgenden Fachausdrücke bezeichnen analoge Begriffe: mittlere Änderungsrate, Differenzenquotient, Durchschnittsgeschwindigkeit

DC2_4_d: Die folgenden Fachausdrücke bezeichnen analoge Begriffe: lokale Änderungsrate, Differentialquotient, Momentangeschwindigkeit

DC2_5_d: Andere Ausdrücke für f'(x): $\frac{df}{dx}$ „df nach dx" (Merziger & Wirth 2010: 13), Tangensteigungsfunktion

DC2_6_e: Falls y=*f*(x), dann kann man *dy*/*dx* entweder interpretieren als Steigungder Kurve *y*=*f*(*x*) oder als Veränderungsrate von *y* bezüglich *x* (Boas 1983: 145, kursiv im Original, übersetzt von SK).

DC3: Derivatives 3 / Ableitungen 3

DC3_1_e: g(x) denotes the function which is the tangent line to the curve y=f(x) at the point (-2,30/28). The tangent line has the equation g(x)=m·x+t with m=f'(-2). "g of x equals m times x plus t with m equals f prime of negative 2."

DC2_2_e: m=f'(-2): The derivative of f at -2, f'(-2), is the slope m of the tangent line at the point (-2,30/28). "m equals f prime of negative 2" or "f prime of negative 2 is the slope of the tangent line to the curve y equals f of x at x equals negative 2". [The expressions used here are no direct quotations, but they follow the wording in the source consulted (Stewart, Redlin & Watson 2002: 924-931).]

DC3_3_e: The equation of the tangent line, g(x), can be obtained by starting with "the point-slope form for the equation of a line through the point (x_1,y_1) with slope m [...] [:] $y - y_1 = m\cdot(x - x_1)$" (Stewart, Redlin & Watson 2002: 925). m = f'(-2) =9/28, x_1 = -2, y_2 =3 0/28

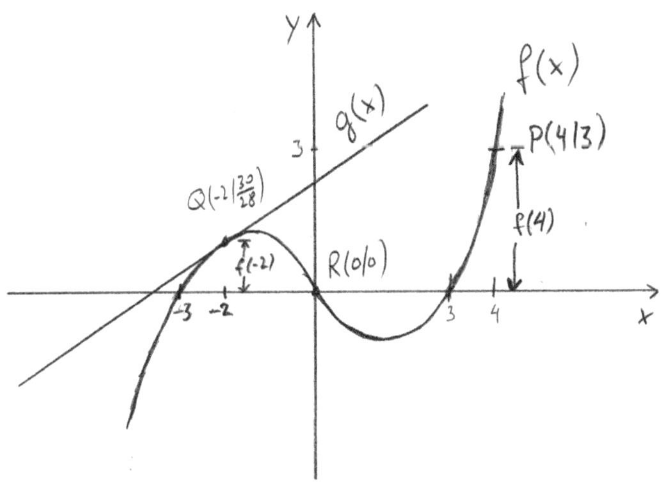

$$f(x)=\frac{3}{28}\cdot\left(x^3-9x\right)$$

$$f'(x)=\frac{9}{28}\cdot(x^2-3)$$

$$y-\frac{30}{28}=f'(-2)\cdot\left(x-(-2)\right)$$

$$y-\frac{30}{28}=\frac{9}{28}\cdot(x+2)$$

$$g(x)=\frac{9}{28}\cdot x+\frac{48}{28}$$

Figure: Graph of the function f(x) and one of its tangent lines, g(x)

Abbildung: Graph der Funktion f(x) und einer ihrer Tangenten, g(x)

DC3_1_d: g(x) bezeichnet die Funktion, welche die Tangente an die Kurve y=f(x) im Punkt Q(-2|30/28) bildet. Die Tangente hat die Gleichung g(x)= m·x+t, m=f'(-2). „g von x gleich m mal x plus t mit m gleich f strich von minus 2."

DC3_2_d: m=f'(-2): Die Ableitung von f bei -2, f'(-2), ist die Steigung m de Tangente am Punkt Q(-2|30/28). „m gleich f Strich von minus 2" oder "f Strich von minus 2 ist die Steigung m der Tangente an die Kurve mit der Gleichung y gleich f von x bei x gleich minus 2".

DC3_3_d: Die Gleichung der Tangente, g(x), kann man erhalten, indem man beginnt mit „der Punkt-Richtungs-Form der Gleichung einer Gerade durch den Punkt $(x_1|y_1)$ mit der Steigung m [...] [:] $y - y_1 = m\cdot(x - x_1)$" (Stewart, Redlin & Watson 2002: 925, übersetzt von SK). m = f'(-2) = 9/28, x_1 = -2, y_2 = 30/28 (siehe oben)

IC1 Areas under Curves 1 / Flächen unter Kurven 1

IC1_1_e: The procedure of integration yields the area between the curve y=f(x) and the x-axis. This area is calculated by using F(x), the indefinite intergral or antiderivative of f(x): “capital F prime of a equals f of x”. C is called constant of integration.

IC2_2_e: The value A of the area between the curve f(x) and the x-axis is „the definite integral with respect to x from a to b of f of x”. This equals “capital F of b minus capital F of a”, which is the value of the antiderivative F at the upper limit b of the intergral minus F at the lower limit a of the integral.

IC2_3_e: If you multiply the estimated mean value of f(x) between a and b, 4, by the difference of the limits of intergration, b-a=2, you get the estimate 8 for the area A.

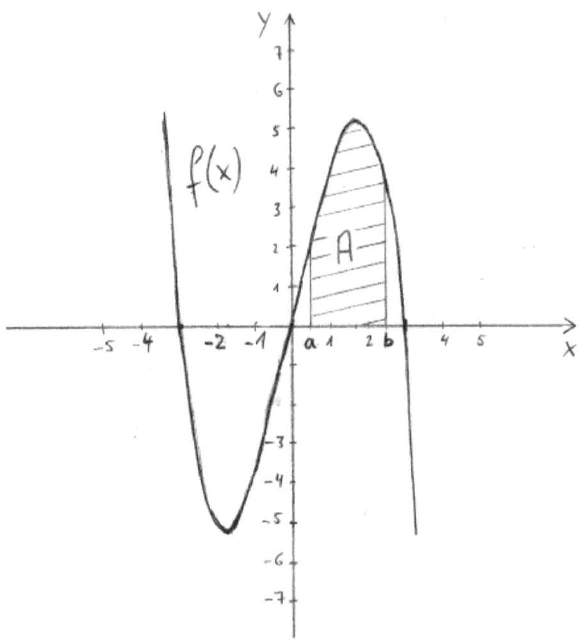

$$f(x) = -\frac{1}{2}\cdot x^3 + \frac{9}{2}\cdot x$$

$$F(x) = -\frac{1}{8}\cdot x^4 + \frac{9}{4}\cdot x^2 + C$$

$$F'(x) = f(x)$$

$$A = \int_a^b f(x)dx = F(b) - F(a)$$

$$A = \int_{1/2}^{5/2} (-\frac{1}{2}\cdot x^3 + \frac{9}{2}\cdot x)dx = \frac{69}{8}$$

$$A = F(\frac{5}{2}) - F(\frac{1}{2}) = \frac{69}{8}$$

$$A \approx 2\cdot 4 = 8$$

Figure: Area under a curve

Abbildung: Fläche unter einer Kurve

IC1_1_d: Das Verfahren der Integration liefert den Inhalt der Fläche zwischen einer Kurve y=f(x) und der x-Achse. Die Fläche wird berechnet mit F(x), dem unbestimmten Intregral oder der Stammfunktion von f(x): „capital F prime of x equals f of x“. C heißt Integrationskonstante.

IC2_2_d: Der Zahlenwert A der Fläche unter dem Graphen von f(x) zwischen den Grenzen a und b ist das „Integral von a bis b f von x d x“. Dies ist gleich „groß F von b minus groß F von a“, also der Wert der Stammfunktion F(x) an der oberen Integrationsgrenze minus dem Wert derStammfunktion an der unteren Integrationsgrenze a.

IC2_3_d: Multipliziert man die Differenz der Integrationsgrenzen, b-a=2, mit dem geschätzten mittleren Funktionswert von f(x) zwischen a und b, 4, erhält man den Schätzwert 8 für die Größe des Flächeninhaltes.

MS1 Modelling Using a Sine Curve 1 / Modellieren mit einer Sinuskurve 1

MS1_1_e: "[A] function *f* is periodic if there is a positive number *p* such that $f(t+p)=f(t)$ for every t, and the least such *p* (if it exists) is the period of *f*"(Stewart, Redlin & Watson 2002: 441, italics in the source). "

MS1_2_3: a>0 is called amplitude; c>0 is called phase shift, in this case a shift to the right; p is the period, the smallest positive number with the property $f(t+p)=f(t)$; b is the number of periods which fit into a length of $2\cdot\pi$; d is the average value of the function; y_{max} is the maximum value; y_{min} is the minimum value; x is a real number;

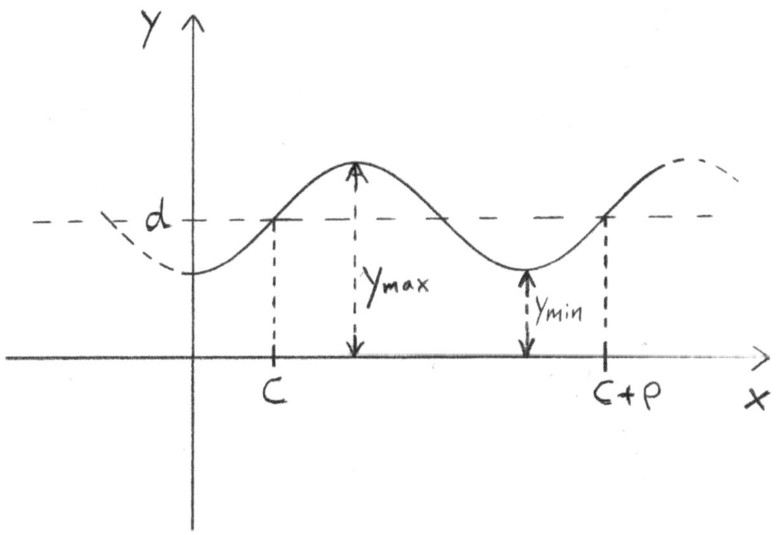

$$d=\frac{y_{max}+y_{min}}{2} \qquad a=\frac{y_{max}-y_{min}}{2} \qquad b=\frac{2\cdot\pi}{p}$$

Figure: Modelling using a Sine Function f(x)=a·sin(b·(x-c)) + d

Abbildung: Modellieren mit einer Sinusfunktion f(x)=a·sin(b·(x-c)) + d

---

MS1_1_d: „Eine Funktion *f* heißt periodisch, wenn für alle $x \in X$ und $x+k\cdot a \in X$ $f(x+k\cdot a)=f(x)$ ($k=0,\pm1,\pm2,...$) gilt [...] Die Konstante *a* heißt Periode der Funktion. Die kleinste Periode *p*, für die $f(x+k\cdot p)=f(x)$ gilt, wird primitive Periode genannt" (Leupold e.a. 1976: 60, kursiv im Original, Fettdruck entfernt).

MS1_2_d: a>0 ist die Amplitude; c>0 wird Phasenverschiebung genannt, in diesem Fall eine Verschiebung nach rechts; p ist die Periode, die kleinste positive Zahl mit der Eigenschaft $f(t+p)=f(t)$; b ist die Zahl der Perioden, die in eine Länge von $2\cdot\pi$ passen; d ist der Durchschnittswert der Funktion; y_{max} ist der Maximalwert; y_{min} ist der Minimalwert; x ist eine reelle Zahl;

CP1: Classical Probability / Klassische Wahrscheinlichkeit

CP1_1_e: "If there are several equally likely, mutually exclusive, and collectively exhaustive outcomes of an experiment, the probability of an event *E* is
p = number of outcomes favourable to *E* / total number of outcomes" (Boas 1983: 686, italics in the source).

CP1_2_e: The probability defined above is sometimes called Laplace probability.

CP1_3_e: "A three digit number (that is, a number from 100 to 999) is selected \`at random.´ [...] What is the probability that all three digits are the same? There are 900 three-digit numbers; 9 of them (namely 111, 222, ..., 999) have all three digits the same. Hence the desired probability is 9/999=1/100" (Boas 1983: 686).

CP1_1_d: „Falls es einige gleich wahrscheinliche, sich gegenseitig ausschließende und in ihrer Gesamtheit vollständige Ergebnisse eines Experimentes gibt, ist die Wahrscheinlichkeit eines Ereignisses *E*
p = Anzahl der für *E* günstigen Ergebnisse / Gesamtzahl der Ergebnisse" (Boas 1983: 686, Kursivschrift im Original, übersetzt von SK).

CP1_2_d: Die oben definierte Wahrscheinlichkeit wird manchmal Laplace-Wahrschein-lichkeit genannt.

CP1_3_d: „Eine dreistellige Zahl (das heißt, eine Zahl zwischen 100 und 999) wird \`zufällig´ ausgewählt. [...] Wie groß ist die Wahrscheinlichkeit, dass alle drei Ziffern gleich sind? Es gibt 900 dreistellige Zahlen, von denen 9 (nämlich 111, 222, ..., 999) drei gleiche Ziffern haben. Daher beträgt die gewünschte Wahrcheinlichkeit 9/999=1/100" (Boas 1983: 686, übersetzt von SK).

TD1: Throwing, Probabilities 1 / Würfeln, Wahrscheinlichkeiten 1

TD1_1_e: "A traditional fair die is thrown twice? What is the probability that: [...] a six turns up exactly once [...] both numbers are odd [...] the sum of the scores is 4?" (Grimmet & Stirzaker 2007: 4)

TD1_2_e: "A red die and a white die are rolled, and the numbers showing are recorded. How many different outcomes are possible? (The singular form of *dice* is *die*.)" (Stewart, Redlin & Watson 2002: 870, italics in the source).

TD1_3_e: "What happens if you roll a die five times? In each independent roll, the probability is 1/6 of rolling a 2 [...] and 5/6 of rolling a different number. In five rolls, the probability of rolling a 2 each time is [...] $(1/6)^5$ [...]. The probability of not rolling a 2 in any roll is [...] $(5/6)^5$" (Herman 2020: 180).

TD1_4_e: "The probability of rolling a 2 in the first four rolls but not in the fifth is [...] $(1/6)^4(5/6)$" (Herman 2020: 180).

TD1_1_d: „Ein herkömmlicher, ungezinkter Würfel wird zweimal geworfen. Wie groß ist die Wahrscheinlichkeit dafür: [...] eine Sechs erscheint genau einmal oben [...] beide Zahlen sind ungerade [...] die Summe der Augenzahlen ist 4?" (Grimmet & Stirzaker 2007: 4).

TD1_2_d: „Ein roter Würfel und ein weißer Würfel werden geworfen und die sich zeigenden Zahlen werden notiert. Wie viele verschiedene Ergebnisse sind möglich? (Der Singular von *dice* [mehrere *Würfel*] ist *die* [ein *Würfel*].)" (Stewart, Redlin & Watson 2002: 870, Kursivschrift in der Quelle, übersetzt von SK).

TD1_3_d: „Was passiert, wenn man eine Würfel fünfmal wirft? Bei jedem unabhängigen Wurf beträgt die Wahrscheinlichkeit, eine 2 zu würfeln [...], 1/6 und 5/6, eine andere Zahl zu würfeln. Bei fünf Würfen ist die Wahrscheinlichkeit, jedes Mal eine 2 zu würfeln [...] $/1/6)^5$ [...]. Die Wahrscheinklichkeit, keine 2 bei irgendeinem Wurf zu erhalten, beträgt [...] $(5/6)^5$"(Hermann 2020: 180, übersetzt von SK).

TD1_4_d: „Die Wahrscheinlichkeit, in den ersten vier Würfen aber nicht im fünften eine 2 zu erhalten, ist [...] $(1/6)^4(5/6)$" (Herman 2020: 180, übersetzt von SK).

OD1: Odds 1 / Wettchancen 1

OD1_1_e: Odds are „[t]he probability that an event will occur divided by the probability that it will *not* occur. If the *probability* of an event occurring is *P*, the odds of the event occurring are equal to *P*/(1-*P*)". [...] Whereas the *probability* of drawing a diamond from a deck of cards is one in four (13/52=1/4=0.25), the *odds* of drawing a diamond are are one in three [0.25/(1-0.25)]=1/3=0.33"(Lang & Secic 1997: 271), italics in the source, boldface removed). In short, 13 of the cards are diamonds and 39=52-13 are not diamonds. Thus, the odds of drawing a diamond are 13/39=1/3.

OD1_2_e: "Probabilities are often stated as odds by giving a ratio, e.g. 9 to 12; this ratio corrresponds to the probability P=9/(9+12)=0.429 (Hedderich & Sachs 2012: 139, boldface removed, translated by SK).

OD1_3_e: "In most gambling games it is customary to express the idea of probability in terms of \`odds against winning´. This is simply the ratio of the unfavourable possibilities to the favourable ones. [...] The probability of getting heads in a toss of a coin is one-half; the odds are 1 to 1" (Glimne 2010).

OD1_4_e: In odds betting, a bookmaker "assesses the probability of victory: 2 to 1, 5 to 1, 1 to 4, and so forth. With a $1 wager on a 2-to-1 underdog, for example, a bettor stands to pocket $2 if the underdog wins. A [...] five dollar bet on a 2-to-5 favourite yields [...] $2" (Thompson 2010).

OD1_5_e: "Most people don't think in terms of odds ratios, nor do they understand what the term means" (Best 2001: 85).

OD1_1_d: Odds [Wettchancen] sind " „[d]ie Wahrscheinlichkeit, dass ein Ereignis eintreten wird, geteilt durch die Wahrscheinlichkleit, dass es *nicht* eintreten wird. Falls *P* die *Wahrscheinlichkeit* für das Eintreten eines Ereignises ist, sind die Odds [Wettchancen] für das Eintreten des Ereignisses gleich *P*/(1-*P*). [...] Während die *Wahrscheinlichkeit*, ein Karo aus einem Stapel Spielkarten zu ziehen, eins zu vier ist (13/52=1/4=0,25), sind die *Odds*, ein Karo zu ziehen, eins zu drei [0,25/(1-0,25)]=1/3=0,33"(Lang & Secic 1997: 271), Kursivschrift im Original, Fettdruck entfernt, übersetzt von SK). Kurz gesagt, 13 der Spielkarten sind Karos und 39=52-13 sind keine Karos. Daher sind die Odds [Wettchancen], ein Karo zu ziehen, 13/39=1/3.

OD1_2_d: „Wahrscheinlichkeiten werden häufig auch als Wettchanchen (odds) durch ein Verhältnis angegeben, z.B. als 9 zu 12; diesem Verhältnis entspricht die Wahrscheinlichkeit P=9/(9+12)=0,429 (Hedderich & Sachs 2012: 139, Fettdruck eintfernt).

OD1_3_d: „In den meisten Glücksspielen ist es Sitte, die Idee der Wahrscheinlichkeit als \`Chance des Verlustes´ anzugeben. Diese ist einfach das Verhältnis der ungünstigen Möglichkeiten zu den günstigen. [...] Die Wahrscheinlichkeit, beim Münzwurf Kopf zu erhalten, ist 1/2, die Wettchance ist 1/1 (Glimne 2010: 104, übersetzt von SK).

OD1_4_d: Bei der Odds-Wette schätzt ein Buchmacher „die Siegeswahrscheinlichkeit: 2 zu 1, 1 zu 4, und so weiter. Mit eine Wette von 1 $ auf einen 2-zu-1-Unterlegenen kann der Tipper 2 $ einnehmen, wenn der Unterlegene gewinnt. Eine Wette [...] von 5 $ auf einen 2-zu-5-Favoriten ergibt [...] 2 $" (Thompson 2010, übersetzt von SK).

OD1_5_d: „Die meisten Leute denken nicht in Quotienten von Wettchancen [Odds] und verstehen auch nicht, was der Ausdruck bedeutet" (Best 2001: 85, übers. von SK).

D Empfohlene Literatur und Quellen / Recommended Literature and Sources

Acton, John R.; Squire, Patrick T. (1985): Solving equations with physical understanding. Bristol: Adam Hilger Ltd

Adams, Dany Spencer (2003): Lab Math. A Handbook of Measurements, Calculations, and Other Quantitative Skills for Use at The Bench. Cold Spring Harbor, NY: Cold Spring Harbor Laboratory Press

Adler, Frederick R. (1998): Modelling the Dynamics of Life. Calculus and Probability for Life Scientists. Pacific Grove, CA: Brooks/Cole

Apostol, Tom M. (1967): Calculus. Volume I. One-Variable Calculus, with an Introduction to Linear Algebra. Second Edition. Waltham, MA: Blaisdell Publishing Company

Arens, Tilo; Busam, Rolf; Hettlich, Frank; Karpfinger, Christian; Stachel, Hellmuth (2022): Grundwissen Mathematikstudium. Analysis und Lineare Algebra mit Querverbindungen.Mit Beiträgen von Klaus Lichtenegger. 2. Auflage. Berlin: Springer Spektrum. [Der Abschnitt „Symbolglossar", Seite 1153-1170, liefert die Aussprache der im Buch verwendteten Symbole auf Deutsch und Englisch zusammen mit einem Hinweis auf die Seite, auf welcher ein Symbol eingeführt wird. Das Buch richtet sich an Studierende der Mathematik, nicht an bloße Anweder. / The section „Symbolglossar", pp. 1153-1170, provides the English and German pronunciation of the symbols used in the book including the number of the page a symbol is introduced on. The book is written for university students of mathematics, not for those primarily interested in applications of mathematics.]

Arens, Tilo; Hettlich, Frank; Karpfinger, Christian; Kockelkorn, Ulrich; Lichtenegger, Klaus; Stachel, Hellmuth (2022): Mathematik, 5, Auflage. Heidelberg: Spektrum Akademischer Verlag. [Der Abschnitt „Symbolglossar", Seite 1645-1653], liefert die Aussprache der im Buch verwendteten Symbole auf Deutsch zusammen mit einem Hinweis auf die Seite, auf welcher ein Symbol eingeführt wird. Das Buch richtet sich an Studierende der Natur- und Ingenieurwissenschaften. / The section „Symbolglossar", pp. 1645-1653, provides the German pronunciation of the symbols used in the book including the number of the page a symbol is introduced on. The book is written for university students of engineering and natural sciences.]

Ayres, Jr., Frank (1963): Schaum's Outline of Theory and Problems of Mathematics of Finance. New York: McGrawHill

Barboianu, Catalin (2022): Understanding the Odds. American Scientist, Vol. 110, March-April, pp. 88-92

Beck, Matthias; Geoghegan, Ross (2010): The Art of Proof. Basic Training for Deeper Mathematics. New York: Springer

Best, Joel (2001): Damned Lies and Statistics. Berkeley: University of California Press

Beller, Sieghard (2004): Empirisch forschen lernen. Konzepte, Methoden, Fallbeispiele, Tipps. Bern: Verlag Hans Huber

Boas, Mary L. (1983): Mathematical Methods in the Physical Sciences. Second Edition. New York: Wiley

Boas, Mary L. (1984): Solutions of Selected Problems for Mathematical Methods in the Physical Sciences. Second Edition. New York: Wiley

Boas, Mary L. (2006): Mathematical Methods in the Physical Sciences. Third Edition. Hoboken, NJ: Wiley

Bormann, Manfred (1978): Experimentalphysik Bd. 1/a. Mechanik der Massenpunkte und des starren Körpers. Manuskript einer Vorlesung für Studenten der Physik im Grundstudium. 4. Auflage 1978. Bochum: Studienverlag Dr. N. Brockmeyer [This book was recommended to me by my physics teacher in 9th grade.]

Bronshtein, Ilya Nikolaevich; Semendyayev, Konstantin Adolfovic (1973): A Guide Book to Mathematics. Translated from the Russian on basis of the enlarged and improved German edition by Jan Jaworowski, Professor at the University of Warsaw, and Michael N. Bleicher, Professor at the University of Wisconsin. New York: Springer Verlag

Bruns, Heinrich (1903): Grundlinien des wissenschaftlichen Rechnens. Leipzig: B.G. Teubner

Burton, Richard F. (1994): Physiology by Numbers. Cambridge: Cambridge University Press

Campbell, Michael T.; Machin, David; Walters, Stephen J. (2007): Medical Statistics. A Textbook for the Health Sciences. Hoboken, NJ: John Wiley & Sons [This book was recommended to me by Michael T. Campbell in 2007]

Causey, Robert L. (1999): magnitude. In: Robert Audi (Ed.): The Cambridge Dictionary of Philosophy. Second Edition. Cambridge: Cambridge University Press. pp. 528-529

Conte, Samuel D.; de Boor, Carl (1981): Elementary Numerical Analysis. An Algorithmic Approach. International Student Edition. 5th Printing 1982. McGraw-Hill Kogakusha Ltd

Dorfmüller, Thomas; Hering, Wilhelm T.; Stierstadt, Klaus; Fischer, Günther (1998): Bergmann·Schäfer. Lehrbuch der Experimentalphysik. Band 1. Mechanik·Relativität·Wärme. 11., völlig neubearbeitet Auflage. Berlin: Walter de Gruyter

Daepp, Ulrich; Gorkin, Pamela (2011): Reading, Writing, and Proving. A Closer Look at Mathematics. Second Edition. New York: Springer

Drucker, C. (Hrsg.) (1925): Ostwald-Luther. Hand- und Hilfsbuch zur Ausführung physikochemischer Messungen. Vierte neubearbeitete Auflage. Leipzig: Akademische Verlagsgesellschaft

Eisenhart, Churchill (1952): The Reliability of Measured Values - Part I. Fundamental Concepts. Photogrammetric Engineering, Vol. XVIII (June), pp. 542-554

Esty, Warren W. (1997): Precalculus Concepts. Preliminary Edition. Upper Saddle River, NJ: Prentice-Hall

Esty, Warren W. (2000): The Language of Mathematics. Sixteenth Edition. Including Chapter 6. Bozeman, MT: Montana State University.

Esty, Warren W. (2012): Precalculus. Sixth Edition. Boston, MA: Pearson

Faires, J. Douglas; Burden, Richard L. (1993): Numerical Methods. Boston: PWS Publishing Company

Gelman, Andrew; Nolan, Deborah (2006): Teaching Statistics. A Bag of Tricks. Oxford: Oxford University Press

Giancoli, Douglas C. (2023): Physics for Scientists and Engineers with Modern Physics. Global Edition. 5th Edition. Pearson: Harlow, Essex

Glimne, Dan (2010): gambling. In: Dale H. Hoiberg & Michael Levy (Eds.): The New Encyclopædia Britannica. 15th Edition. Micropædia. Volume 5. Chicago: Encyclopædia Britannica, Inc. p. 104-105 [The name of the author is stated in the DVD-ROM edition.]

Gould, Martin; Hurst, Edward (2009): Bridging the Gap to University Mathematics. London: Springer-Verlag

Hedderich, Jürgen; Sachs, Lothar (2012): Angewandte Statistik. Methodensammlung mit R. Vierzehnte, überarbeitete und ergänzte Auflage. SpringerGabler

Henderson, Michael (1990): The BMA Guide to Living with Risk. The British Medical Association. London: Penguin

Herman, Irving P. (2020): Coming Home to Math. Become Comfortable with the Numbers that Rule Your Life. Singapore: World Scientific [This book is full of applications of mathematics to the everyday world.]

Hertz, Heinrich (1894): Die Prinzipien der Mechanik. Gesammelte Werke. Band 3. Leipzig: Barth

Hertz, Heinrich (1899): Principles of Mechanics. Collected Works. Third Volume. Authorized English Translation By D. E. Jones, BSc., And J.T. Walley, M.A. London: Macmillan

Jacques, Ian (2006): Mathematics for Economics and Business. Fifth Edition. Harlow, England: Pearson Education Ltd.

Grimmett, Geoffrey R.; Stirzaker, David R. (2007): One Thousand Exercices in Probability. Oxford: Oxford University Press

Hoiberg, Dale H.; Levy, Michal (Eds.) (2010): "error" In: The New Encyclopædia Britannica. 15th Edition. Micropædia. Volume 4. Chicago: Encyclopædia Britannica, Inc. p. 548

Kern, Wolfgang (1993): Sie halten sich für mittelmäßig, sind jedoch viel schlechter. FU:N 2/93. Berlin: Freie Universität Berlin

Kelly, Gerard W. (1984): Short-Cut Math. New York: Dover Publications

Kirkup, Les (1994): Experimental Methods. An Introduction to the Analyis and Presentation of Data. Brisbane: John Wily & Sons

Knorrenschild, Michael (2013): Numerische Mathematik. Eine beispielorientierte Einführung. Leipzig: Fachbuchverlag Leipzig im Carl Hanser Verlag

Kreul, Hans; Ziebarth, Harald (2016): Mathematik leicht gemacht. 8., überarbeitete und erweiterte Auflage. 42781 Haan-Gruiten: Verlag Europa-Lehrmittel. [Dieses Buch hilft Ihnen dabei, den Stoff der deutschen Schulmathematik von Klasse 5 bis Klasse 10 zu lernen. Es enthält ausführliche Erklärungen von Begriffen und Vorgehensweisen, durchgerechnete Beispiele und Hunderte von Übungsaufgeben mit Lösungen. / This book helps you to learn the subject matter of German school mathematics from 5th grade to 10th grade. It contains extensive explanations of concepts and procedures, worked examples and hundreds of problems with solutions.]

Lang, Thomas A.; Secic, Michelle (1997): How To Report Statistics In Medicine. Annotated Guidelines for Authors, Editors, and Reviewers. Philadelphie: American College of Physicians

Läuger, Klaus (1992): Mathematik kompakt ... und verständlich. Für das naturwissenschaftliche Grundstudium an Hochschulen und Fachhochschulen. Insbesondere für Studienanfänger mit Haupt- und Nebenfach Physik. 2. Auflage. München: R. Oldenbourg Verlag

Leupold, Wilhelm; Conrad, Rudolf; Völkel, Siegfried; Große, Gerhard; Fucke, Rudolf; Nickel, Heinz; Mende, Horst (1976): Analysis für Ingenieure. 12. Auflage. Zürich: Verlag Harri Deutsch [Dieses Buch erklärt die deutsche Schulmathematik der Klassen 11 bis 13 und einige Grundlagen der Ingenieurmathematik der ersten zwei Semster.]

Lübsen, Heinrich Borchert (1869): Ausführliches Lehrbuch der Arithmetik und Algebra zum Selbstunterricht und mit Rücksicht auf die Zwecke des practischen Lebens. Elfte unveränderte Auflage. Leipzig: Friedrich Brandstetter. Digital Archive of the Bayerische Staatsbibliothek, Munich. Math. P. 705. urn:nbn:de:bvb:12-bsb11011242-6. [Mit diesem Buch lernte Albert Einstein Mathematik. / This is the book Albert Einstein learnt mathematics with.]

Lübsen, Heinrich Borchert (1897): Mathematics Self-Taught. The Lübsen Method for Self-Instruction, and Use in the Problems of Practical Life; Arithmetic and Algebra. Adatd from the German of H. B. Lübsen by Henry Harrison Suplee. London: FB &c Ltd.

Maddox, John (1972): The Doomsday Syndrome. London: Macmillan

Maddox, John (1992): Language for a polyglot readership. Nature, Vol. 359, Issue No. 6395, p. 475, 18 October 1992

Mahajan, Sanjoy (2018): Decluttering our thinking with the life-changing magic of twiddle. American Journal of Physics, Vol. 86 (2), 143-145

Mahajan, Sanjoy (2022): Approximate insightful ODE solutions. American Journal of Physics, Vol. 90 (12), 887-888

Malone, Leo J.; Dolter, Theodore O.; Gentemann, Stephen (2013): Basic Chemistry. 9th Edition. Singapore: John Wiley & Sons

McSeveny, Alan; Conway, Rob; Wilkes, Steve, Smith, Michael (2014): International Mathematics for the Middle Years 1. Melbourne, Victoria: Pearson Australia

Mende, Dietmar; Simon, Günter (2016): Physik. Gleichungen und Tabellen. 17., aktualisierte Auflage. München: Carl Hanser Verlag

Merziger, Gerhard; Wirth, Thomas (2010): Repetitorium Höhere Mathemtik. 6. Auflage. Barsinghausen: Binomi Verlag

Meyberg, Kurt; Vachenauer, Peter (2003): Höhere Mathematik 1. Differential- und Integralrechnung. Vektor- und Matrizenrechnung. 6. Auflage. Berlin: Springer

Munroe, M. Evans (1963): The Language of Mathematics. Ann Arbor: The University of Michigan Press

Nachtigall, Christof; Wirtz, Markus (1998): Wahrscheinlichkeitsrechnungund Inferenzstatistik. Statistische Methoden für Psychologen. Teil 2. Weinheim: Juventa

Olive, Jenny (2003): Maths. A Student´s Survical Guide. Cambridge: Cambridge University Press

Paulos, John Allen (1990): Zahlenblind. Mathematisches Analphabetentum und seine Konsequenzen. Mit einem Vorwort von Douglas R. Hofstadter- München: Wilhelm Heyne Verlag

Paulos, John Allen (2001): Innumeracy. Mathematical Illiteracy and Its Consequences. New York: Hill and Wang

Pfeifer, Andreas (2006): Praktische Finanzmathematik. Mit Futures, Optionen, Swaps und anderen Derivate. Frankfurt am Main: Verlag Harri Deutsch

Phillips, John L. (2000): How to Think About Statistics. Sixth Edition. New York: W. H. Freeman

Polya, George; Kilpatrick, Jeremy (1974): The Stanford Mathematics Problem Book. With Hints and Solutions. New York: Teachers College Press

Poole, David (2003): Linear Algebra. A modern Introduction. Pacific Grove, CA: Brooks/Cole

Riley, Kenneth F. (1982): Problems for Physics Students. Cambridge: Cambridge University Press

Riley, Kenneth F.; Hobson, Michael P.; Bence, Stephen J. (2006): Mathematical Methods for Physics and Engineering. Third Edition. Cambridge: Cambridge University Press

Rogers, Louis; Willoughby, Dawn (2013): Numbers. Data and statistics for the non-specialist. London: HarperCollins Publishers

Romer, Robert H. (1976): Energy. An Introduction to Physics. San Francisco: W. H. Freeman and Company

Schick, Karl (1976): Lineare Optimierung. Mannheim: BI-Wissenschaftsverlag

Scholz, Günter; Vogelsang, Klaus (1991): Kleines Lexikon : Einheiten, Formelzeichen, Größen. Leipzig: Fachbuchverlag Leipzig

Stevenson, Angus; Waite, Maurice (Eds.) (2011): Concise Oxford English Dictionary. Twelfth Edition. Oxford: Oxford University Press

Stewart, James; Redlin, Lother; Watson, Saleem (2002): Precalculus. Mathematics for Calculus. 4th Edition. Pacific Grove, CA: Brooks/Cole [This book can help you to survive school mathematics from 7th to 12th grade.]

Taylor, John R. (1982): An Introduction to Error Analysis. The Study of Uncertainties in Physical Measurements. Mill Valley, CA: University Science Books

Taylor, John R. (1988): Fehleranalyse. Eine Einführung in die Untersuchung von Unsicherheiten in physikalischen Messungen. Weinheim: VCH

Thiele, Rüdiger (1979): Mathematische Beweise. Thun: Verlag Harri Deutsch; Leipzig: BSB B. G. Teubner Verlagsgesellschaft

Topping, James (1963): Errors of Observation and their Treatment. Third Edition. London: Chapman and Hall

Topping, James (1975): Fehlerrechnung. Übersetzt von Jochen Schwarze unter Mitarbeit von Peter Erven. Weinheim: Physik Verlag

UK, Office for National Statistics (2016): United Kingdom Annual Abstract of Statistics, No. 152. London: Dandy Booksellers

Weil, Jonathan F. (2007): Units of measurement. In: Elizabeth Geller & Jonathan Weil (Eds.): McGraw-Hill Encyclopedia of Science & Technology. 10th Edition. Volume 19. pp. 70-77

Weinstein, Lawrence (2012): Guesstimation 2.0. Solving Today's Problems on the Back of a Napkin. Princeton: Princeton University Press.

Weinstein, Lawrence & Adam, John A. (2008): Guesstimation: solving the world´s problems on the back of a cocktail napkin. Princeton: Princeton University Press.

E Dictionaries of Mathematics / Wörterbücher der Mathematik

Berry, John; Graham, Ted; Sharp, Jenny; Berry, Elizabeth (2000): The complete A-Z mathematics handbook. Second Edition. For AS and A level. Abingdon, Oxon: Hodder & Stoughton

Borowsky, Ephraim J.; Borwein, Jonathan M. (1989): Collins dictionary of mathematics. Reprint 10. HarperCollins

Eisenreich, Günther; Sube, Ralf (1982): Dictionary of Mathematics. English. German, French, Russian. Containing approximately 35000 terms. Volume I. A-Z. Wörterbuch Mathematik, Englisch. Deutsch. Französisch. Russisch. Mit etwa 35000 Wortstellen. Band I. A-Z. Thun: Verlag Harri Deutsch

Eisenreich, Günther; Sube, Ralf (1982): Dictionary of Mathematics. English. German, French, Russian. Containing approximately 35000 terms. Volume II. [Index]. Wörterbuch Mathematik, Englisch. Deutsch. Französisch. Russisch. Mit etwa 35000 Wortstellen. Band II. Register. A-Z. Thun: Verlag Harri Deutsch

Harrison, Ian (1998): The complete A-Z Accounting handbook. Abingdon, Oxon: Hodder & Stoughton

Herland, Leo (1965): Dictionary of Mathematical Sciences. Volume I. German-English. Second Edition. Revised and Enlarged. Wörterbuch der mathematischen Wissenschaften. Band I. Deutsch-Englisch. Zweite, erweiterte und verbesserte Auflage. New York: Frederick Ungar Publishing Co.

Herland, Leo (1965): Wörterbuch der mathematischen Wissenschaften. Volume II. Englisch-Deutsch. Mit einem Supplement von neuen Wörtern. Dictionary of Mathematical Sciences. Volume II. English-German. With a Supplement of New Words. New York: Frederick Ungar Publishing Co.

James, Robert C.; Alchian, Armen A.; Beckenbach, Edwin F.; Bell, Clifford; Craig, Homer V.; James, Glenn; Michal, Aristotle D.; Sokolnikoff, Ivan S. (1992): Mathematics Dictionary. Fifth Edition. New York: Chapman & Hall [Das Buch enthält eine Liste mathematischer Symbole (Seite 455-460), teilweise mit Aussprache auf Englisch, und ein Register (Seite 478-548) auf Französisch, Deutsch, Russisch und Spanisch. / The book contains a list of mathematical symbols (pp. 455-460), some with their English pronunciation, and an index in French, German, Russian, and Spanish.]

Lewisch, Ingrid; Posamentier, Alfred S. (2000): Mathematisches Fachwörterbuch. Englisch-Deutsch. Deutsch-Englisch. 2. Auflage. Wien: R. Oldenbourg Verlag

Rogers, Kirsten; Large, Lori (2012): Usborne Junior Illustrated Maths Dictionary. Designed by Karen Tomlins. Illustrated by Ruth Russell. London: Usborne Publishing Ltd.

Städtler, Ingrid; Städtler, Werner (1966): Englisch. Symbolik und Fachausdrücke. Mathematik, Physik, Chemie. Mit 43 Bildern. Leipzig: VEB Fachbuchverlag [Die Seiten 8-41 liefern die Aussprache mathematischer Symbole auf Deutsch und Englisch. / The pages 8-41 provide the English and German pronunciation of mathematical symbols.]

Tapson, Frank (1999): The Oxford Mathematics Study Dictionary. Second Edition. Oxford: Oxford University Press

If you need to know the German word for an English word denoting a concept you are familiar with, e.g., "derivative - Ableitung", some books from this list of dictionaries will help you (Städtler & Städtler 1966: 25; Herland 1965: Volume II; Eisenreich & Sube 1982; Lewisch & Posamentier 2000: 33). In school mathematics from fifth to tenth grade, the book by Tapson serves that purpose, e.g. in the case of "prime number - Primzahl" (Tapson 1999: 169). If you need the definition of a concept, e.g. "derivative", you can consult dictionaries in English (Borowski & Borwein 1989: 151; James e.a. 1992: 112; Berry e.a. 2000: 66). From grade 5 to grade 10, three books from my list will help you, e.g., in the case of "prime number" (Tapson 1999: 44; Berry e.a. 2000: 187; Rogers & Large 2012: 13).

Falls Sie das englische Wort für einen mathematischen Begriff suchen, mit dem Sie vertraut sind, ist das „Dictionary of Mathematics" (Eisenreich & Sube 1982) die erste Wahl. Aber auch das „Dictionary of Mathematical Sciences, Volume I" (Herland 1965) leistet noch gute Dienste. Das „Mathematics Dictionary" (James e.a. 1992) liefert Ihnen das Wort und die Begriffserklärung.

F Fun / Spaß

Die Ballade vom idealen Mathematiklehrer

Der ideale Lehrer verdient ein Lob
Verglichen mit ihm sind die anderen grob.
  Es ist ein solcher idealer Mann,
  Der unser aller Zukunft sichern kann.

Er erklärt Mathe, dass es jeder versteht.
Nur bei ihm lernt einjeder, wie etwas geht.
  Es ist ein solcher idealer Mann,
  Der uns alle das Lernen lehren kann.

Und niemand bleibt beim ihm lange im Dunkeln,
So dass alle Schüleraugen hell funkeln.
  Es ist ein solcher idealer Mann,
  Der unser Selbstvertrauen heben kann.

Er hat keine Launen und keine Zoten
Und er vergibt auch nur gerechte Noten.
  Es ist ein solcher idealer Mann,
  Der uns Gerechtigkeitssinn lehren kann.

Seine Schulaufgaben sind niemals zu schwer.
Auch deshalb mögen ihn alle Schüler sehr.
  Es ist ein solcher idealer Mann,
  Der bei uns schweren Frust verhindern kann.

Aber dieser Lehrer hat einen Makel,
Und dieser macht die Schule zum Debakel.
  Denn was einjeder an ihm kritisiert,
  Dass dieser Lehrer gar nicht *existiert*.

Falls Sie oder Ihre Kinder unter dieser Situation leiden und in München und Umgebung wohnen, können Sie Unterrichtstermine mit mir vereinbaren: sixtus.kage@web.de

Ich erteile Unterricht in Mathematik, Physik und anderen mathematiklastigen Fächern (u.a. BWL, Ingenieurwssenschaften, Statistik für Mediziner), und zwar auf Deutsch und Englisch, von der Schule bis zur Universität, einschließlich International Baccalaureate.

[In the second edition of this book, I am going to publish an English version of the ballad.]

G Danksagungen / Acknowledgements

I thank my mathematics teacher at the Jesuit high school in Hamburg, H.-J. K., for his innovative and interesting lessons from grade 7 to grade 10. He and his former mathematics teacher practised a successful method. Their students were induced to learn the verbal descriptions of mathematical methods by heart. He also recommended me to read the book by Bormann (Bormann 1978), which I cite in this book. I also thank my other mathematics teacher at the Jesuit high school, P.-G. P., for the opportunity to take over the teaching in grade 11 to grade 13 for up to two weeks per semester. I thank the mathematics professors Zenger and Sauter (Technical University of Munich) for perfect lectures and tutoring sessions in numerical mathematics. I thank my customers, about 1800 (one thousand eight hundred) students at German-language and English-language schools and universities, for the opportunity of tutoring them in mathematics, physics, engineering and other subjects. The total number of hours has been more than 50000 (fifty thousand) so far. Finally, I thank the staff of the Library of the Deutsches Museum in Munich for providing a perfect service and a pleasant working atmosphere.

Meinem Mathematiklehrer H.-J. K. (Jesuitengymnasium in Hamburg) danke ich für methodisch innovativen und interessanten Unterricht in den Klassen 7 bis 10. Von ihm und seinem früheren Mathematiklehrer stammt eine erfolgreiche Methode. Ihre Schüler wurden veranlasst, die sprachlichen Beschreibungen mathematischer Handlungen auswendig zu lernen. Er empfahl mir auch, das hier zitierte Buch von Bormann (Bormann 1978) zu lesen. Meinem anderen Mathematiklehrer P.-G. P. (Jesuitengymnasium in Hamburg) danke ich für die Möglichkeit, in den Klassen 11 bis 13 bis zu zwei Wochen pro Semester den Unterricht übernehmen zu dürfen. Den Professoren Zenger und Sautter (Technische Universität München) danke ich für perfekte Vorlesungen und Übungen in Numerischer Mathematik. Meinen Kunden, bislang etwa 1800 Schülern, Schülerinnen und Studierenden an deutschsprachigen und englischsprachigen Schulen und Universitäten, danke ich für die Gelegenheit, Erfahrung im Unterrichten von Mathematik, Physik, Ingenieurwissenschaften und anderen Fächern zu sammlen. Es handelt sich um bislang mehr als 50000 (fünfzigtausend) Stunden. Schließlich danke ich dem Personal der Bibliothek des Deutschen Museums in München für perfekten Service und eine angenehme Arbeitsatmosphäre.

H Blurb / Klappentext

In this booklet, you find the language of describing and explaining mathematical concepts, symbols, and procedures in English and German. This booklet is intended to assist people who have to learn, to teach or to apply mathematics while using English or German as a foreign language. The pronunciation of the words and symbols is provided in the the author's YouTube channel, https://www.youtube.com/@SixtusKage, page by page, in English and German videoclips separately. This service is free, is not included in the price of the booklet and can be terminated or modified any time.
In diesem Büchlein findet man die Sprache der Beschreibung und Erklärung mathematischer Begriffe, Symbole und Prozeduren. Dieses Büchlein soll Leute unterstützen, die Mathematik lernen, anwenden oder lehren müssen, während sie Englisch oder Deutsch als Fremdsprache nutzen. Die Aussprache der Wörter und Symbole wird im YouTube-Kanal des Autors geliefert, https://www.youtube.com/@SixtusKage, Seite für Seite in Videofilmen, getrennnt auf Deutsch und auf Englisch. Dieser Service ist gebührenfrei, ist nicht im Preis des Büchleins enthalten und kann jederzeit beendet oder verändert werden.

The author: Sixtus Kage holds university dregrees in physics and psycholinguistics. For decades, he has tutored students preparing for the Abitur, for the International Baccalaurate IB, and for university exams in Germany, England, Switzerland, Italy, and the USA. The author has also provided support in the preparation and evaluation of research projects. Sixtus Kage writes satires and humorous poems in English, German and, occasionally, Latin.
Der Autor: Sixtus Kage hat Universitätsabschlüsse in Physik und Psycholinguistik. Er hat jahrzehntelang Nachilfeunterricht erteilt zur Vorbereitung auf das Abitur, das International Bacclaureate IB und auf Prüfungen an Universitäten in Deutschland, England, Italien, der Schweiz und den USA. Der Autor hat auch Unterstützung bei der Vorbereitung und Auswertung wissenschaftlicher Arbeiten gegeben. Sixtus Kage schreibt Satiren und humoristische Gedichte auf Englisch, Deutsch und gelegentlich Latein.

Das Foto (aufgenommen von Sixtus Kage) zeigt den Autor auf dem Weg zu seiner Lieblingsbibliothek in München.
The photo (taken by Sixtus Kage) shows the author on his way to his favourite library in Munich.